Practice Papers for SQA Exams

Higher

Text © 2009 Billy Dickson and Graham Moffat
Design and layout © 2009 Leckie & Leckie

02/050213

All rights reserved. No part of this publication may be reproduced, stored in a retrieval system, or transmitted in any form or by any means, electronic, mechanical, photocopying, recording or otherwise, without prior permission in writing from Leckie & Leckie Ltd. Legal action will be taken by Leckie & Leckie Ltd against any infringement of our copyright.

The right of Billy Dickson and Graham Moffat to be identified as the authors of this Work has been asserted by them in accordance with sections 77 and 78 of the Copyright, Designs and Patents Act 1988.

ISBN 978-1-84372-779-8

Published by
Leckie & Leckie Ltd
An imprint of HarperCollins*Publishers*
Westerhill Road, Bishopbriggs, Glasgow, G64 2QT
T: 0844 576 8126 F: 0844 576 8131
leckieandleckie@harpercollins.co.uk www.leckieandleckie.co.uk

A CIP Catalogue record for this book is available from the British Library.

Questions and answers in this book do not emanate from SQA. All of our entirely new and original Practice Papers have been written by experienced authors working directly for the publisher.

FSC™ is a non-profit international organisation established to promote the responsible management of the world's forests. Products carrying the FSC label are independently certified to assure consumers that they come from forests that are managed to meet the social, economic and ecological needs of present and future generations, and other controlled sources.

Find out more about HarperCollins and the environment at
www.harpercollins.co.uk/green

Introduction

The three practice exams in this book together give an overall and comprehensive coverage of assessment of **Knowledge** and **Problem Solving** in Higher Biology. The **Topic Index** on page 4 shows the pattern of coverage.

We recommend that candidates download and print a copy of the **Arrangements** for **Higher Biology** from the SQA website www.sqa.org.uk following the links through Services for Learners. Print pages 6 to 35 which show the Knowledge required and also page 38 which shows the Problem Solving skills which **will** be tested.

Using the Papers

Each paper can be attempted in its entirety or groups of questions on a particular topic or skill area can be attempted using the **Topic Grid** for reference. Use the target date column to record progress.

We recommend toggling between doing questions and studying their worked answers.

Each paper has been carefully assembled to be as similar as possible to a typical Higher Biology Paper. Each has three Sections (A, B and C).

Section A contains 30 Multiple Choice items, about 20 of which test **Knowledge** and the remainder test **Problem Solving** skills. The section is for 30 marks.

Section B has short answer questions for 80 marks. There is a ratio of about 2 or 3 marks of **Knowledge** to 1 mark of **Problem Solving** in Section B. The Problem Solving questions include a Data Handling question in which a related set of data is presented and a Practical Situation in which an investigation or experiment is described in some detail.

Section C has two extended response questions with a choice in each. Extended responses test **Knowledge** and understanding. Each question is for 10 marks giving a total of 20 marks for the Section.

Most questions are set at the standard for a C Grade but there are also a few more difficult questions set at the standard for an A Grade. Each paper has been carefully constructed to represent the **typical range of demand** in a Higher Biology Paper.

The Worked Answers

The worked answers on pages 101–157 give acceptable answers with alternatives. Each answer has a helpful tip provided. We have tried to give a variety of helpful tips - hints on the Biology itself including some memory ideas, focus on traditionally difficult areas for candidates, advice on wording answers and general points to watch for.

Grades

The papers are designed to be equally difficult and to reflect the national standard of a typical SQA paper. Each paper has 130 marks – if you score 65 marks that's a **C** pass. You will need at least 80 marks for a **B** pass and about 100 for an **A**. These figures are a rough guide only.

Timing

If you are attempting a full paper, limit yourself to 2 hours and 30 minutes to complete. Get someone to time you! We recommend no more than 30 minutes for Section A, 1 hour and 20 minutes on Section B and 40 minutes on Section C.

Higher Biology Practice Exam Papers

Topic Index Grid

Each Topic Area has questions from each Exam. Find the questions as follows – eg A21 is **Section A** Question 21, eg B8 is **Section B** Question 8, eg C1B is **Section C** Question 1 Option B

Bold Italics indicate data questions, **Bolds** indicate practical situation question.

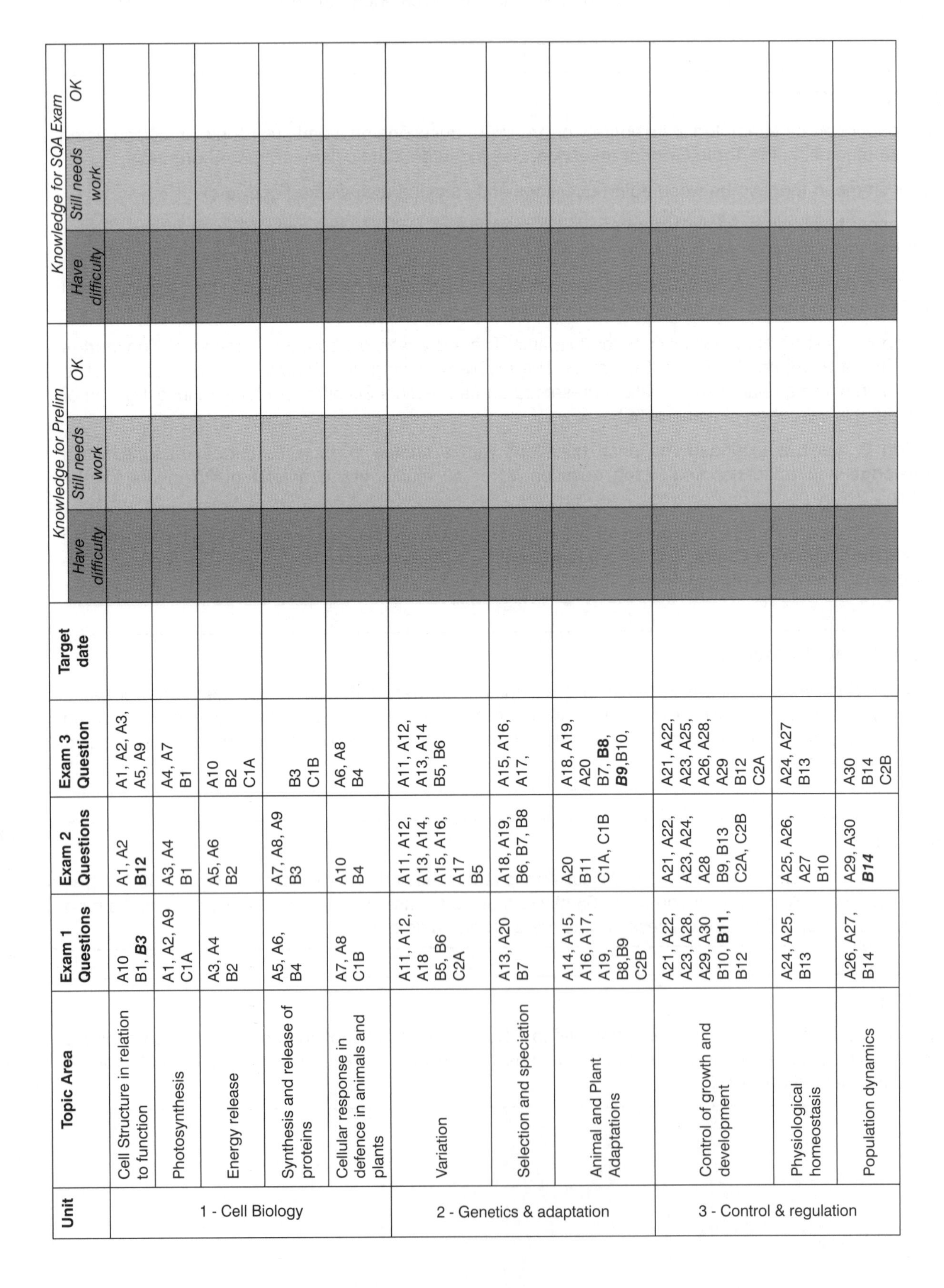

Unit	Topic Area	Exam 1 Questions	Exam 2 Questions	Exam 3 Question	Target date	Knowledge for Prelim			Knowledge for SQA Exam		
						Have difficulty	*Still needs work*	*OK*	*Have difficulty*	*Still needs work*	*OK*
1 - Cell Biology	Cell Structure in relation to function	A10 B1, ***B3***	A1, A2 **B12**	A1, A2, A3, A5, A9							
	Photosynthesis	A1, A2, A9 C1A	A3, A4 B1	A4, A7 B1							
	Energy release	A3, A4 B2	A5, A6 B2	A10 B2 C1A							
	Synthesis and release of proteins	A5, A6, B4	A7, A8, A9 B3	B3 C1B							
	Cellular response in defence in animals and plants	A7, A8 C1B	A10 B4	A6, A8 B4							
2 - Genetics & adaptation	Variation	A11, A12, A18 B5, B6 C2A	A11, A12, A13, A14, A15, A16, A17 B5	A11, A12, A13, A14 B5, B6							
	Selection and speciation	A13, A20 B7	A18, A19, B6, B7, B8	A15, A16, A17,							
	Animal and Plant Adaptations	A14, A15, A16, A17, A19, B8,B9 C2B	A20 B11 C1A, C1B	A18, A19, A20 B7, **B8**, ***B9***,B10,							
3 - Control & regulation	Control of growth and development	A21, A22, A23, A28, A29, A30 B10, **B11**, B12	A21, A22, A23, A24, A28 B9, B13 C2A, C2B	A21, A22, A23, A25, A26, A28, A29 B12 C2A							
	Physiological homeostasis	A24, A25, B13	A25, A26, A27 B10	A24, A27 B13							
	Population dynamics	A26, A27, B14	A29, A30 ***B14***	A30 B14 C2B							

Topic Index Grid (Problem Solving)

Problem Solving		Exam 1 Questions	Exam 2 Questions	Exam 3 Questions	Knowledge for Prelim			Knowledge for SQA Exam		
					Have difficulty	*Still needs work*	*OK*	*Have difficulty*	*Still needs work*	*OK*
Problem Solving A Handling Information	Selecting information	A2, A15 B3ai, B3bi, B3c, B9bi, B13ci, B14ai, B14aii, B14aiii	B4f, B11ai, B13cii, B14a, B14c, B14e	A30 B1ai, B7aii, B8c, B9ai, B9aiii, B9aiv, B10ai, B11ai, B11aii, B11aiii,						
	Presenting information	B11bi, B11bii	B12a	B8b						
	Processing information	A9, A16, A23 B1ai, B3aii, B3d, B4b, B5a, B5b, B9bii, B13cii	A4, A9, A11, A13, A24, A27 B3f, B4c, B5a, B13ci, B14f, B14g	A9, A11, A12, A19, A20, A28, B6b, B7ai, B8bi, B14ai						
Problem Solving B Experimental Procedure	Planning and designing	A28, B11ai, B11aiii	B12di	B8aii, B8aiii, B8aiv						
	Evaluating	B11aii	B12c	B8ai						
	Concluding and predicting	A4, A10, A17,A27, A29 B1aii, B3bii, B11c, B11di, B11dii	A2, A28, A29 B12bi, B12e, B14b, B14d	A10, A29 B6a, B6c, B9aii, B10aii, B10aiii, B12a						

Answer sheet for Section A:

Higher Biology
Practice Papers for SQA Exams

Please select your answer using a single mark e.g. A ☐ B ■ C ☐ D ☐

	A	B	C	D
1.	☐	☐	☐	☐
2.	☐	☐	☐	☐
3.	☐	☐	☐	☐
4.	☐	☐	☐	☐
5.	☐	☐	☐	☐
6.	☐	☐	☐	☐
7.	☐	☐	☐	☐
8.	☐	☐	☐	☐
9.	☐	☐	☐	☐
10.	☐	☐	☐	☐
11.	☐	☐	☐	☐
12.	☐	☐	☐	☐
13.	☐	☐	☐	☐
14.	☐	☐	☐	☐
15.	☐	☐	☐	☐

	A	B	C	D
16.	☐	☐	☐	☐
17.	☐	☐	☐	☐
18.	☐	☐	☐	☐
19.	☐	☐	☐	☐
20.	☐	☐	☐	☐
21.	☐	☐	☐	☐
22.	☐	☐	☐	☐
23.	☐	☐	☐	☐
24.	☐	☐	☐	☐
25.	☐	☐	☐	☐
26.	☐	☐	☐	☐
27.	☐	☐	☐	☐
28.	☐	☐	☐	☐
29.	☐	☐	☐	☐
30.	☐	☐	☐	☐

Exam 1

Biology Higher

Practice Papers
For SQA Exams

Exam 1

You have two hours, 30 minutes to complete this paper.

Try to answer all of the questions in the time allowed.

Write your answers in the spaces provided, including all of your working.

SECTION A

All questions in this section should be attempted.

Answers should be given on the separate answer sheet provided on page 6

1. Which line in the table shows correctly the number of carbon atoms in molecules of substances involved in the Calvin Cycle?

	Number of carbon atoms per molecule		
	Carbon dioxide	*RuBP*	*GP*
A	1	5	3
B	2	3	5
C	2	5	3
D	1	3	5

2. The graph shows how the rate of photosynthesis in a water plant changed as the carbon dioxide concentration in the water in which it was growing was increased. The plant was kept at a constant high light intensity and at a temperature of 10°C.

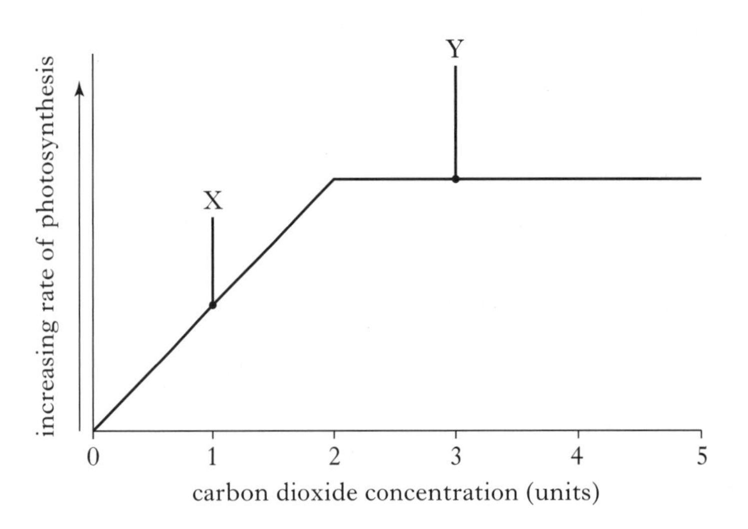

Which line in the table correctly shows the factors which were most likely to have limited the rate of photosynthesis at points X and Y on the graph?

	X	*Y*
A	light intensity	carbon dioxide concentration
B	temperature	carbon dioxide concentration
C	carbon dioxide concentration	light intensity
D	carbon dioxide concentration	temperature

3. Which of the following statements refer to glycolysis?

1 carbon dioxide is released
2 it occurs during aerobic respiration
3 the end product is pyruvic acid
4 the end product is lactic acid

A 1 and 3
B 1 and 4
C 2 and 3
D 2 and 4

4. Three mutant strains of yeast each lacked a different respiratory enzyme and could not carry out specific metabolic reactions as shown below.

Strain P – could not convert pyruvic acid to ethanol
Strain Q – could not attach an acetyl group to CoA
Strain R – could not form citric acid

In which strains could the Krebs cycle still occur?

A P and Q
B P only
C Q and R
D R only

5. Which line in the table shows a correct classification for proteins found in cells?

	Protein found in cells	*Classification*
A	hormone	fibrous
B	collagen	globular
C	enzyme	globular
D	antibody	fibrous

6. The function of tRNA in cell metabolism is to

A carry codons to the ribosomes
B transport amino acids for protein synthesis
C transcribe the DNA code
D synthesise anticodons.

7. The stages of infection of a host cell by a virus are listed below.

1 viral protein synthesised
2 virus binds to host cell
3 viral nucleic acid replicates
4 viral DNA enters host cell

The sequence in which these events occur is

A 2, 4, 3, 1
B 2, 4, 1, 3
C 4, 3, 1, 2
D 4, 1, 3, 2

8. Which substance produced by plants in response to attack by pathogens isolates affected areas of plant tissue from the rest of the plant?

A tannin
B cyanide
C resin
D nicotine

9. A culture of green algae was kept at a constant temperature and subjected to periods of dark and light. It consumed an average of 7 cm^3 of oxygen per hour by respiration in either light or dark conditions. When in light conditions it produced an average of 23 cm^3 of oxygen per hour by photosynthesis.

What is the net volume of oxygen produced by the algae over a 24 hour period consisting of 14 hours of dark followed by 10 hours of light?

A 38 cm^3
B 62 cm^3
C 132 cm^3
D 160 cm^3

10. The table below shows the concentrations of three different ions found in sea water and in the sap of the cells of a seaweed species.

	Ion concentration (mgl^{-1})		
	potassium	sodium	chloride
Sea water	0.05	0.56	0.64
Seaweed cell sap	0.58	0.06	0.61

Which of the following statements can be supported by the data in the table?

A Potassium and sodium ions are taken into the cell by active transport.

B Potassium and chloride ions are removed from the cell by diffusion.

C Sodium ions are removed from the cell by active transport.

D Chloride and sodium ions are removed from the cell by diffusion.

11. Sex-linked alleles in humans **cannot** be passed from

A fathers to their sons
B women to grandsons
C mothers to their sons
D men to their grandsons.

12. The diagrams show a chromosome before and after a chromosome mutation. The numbers refer to the locations of genes on the chromosome.

Before mutation

1	2	3	4	5	6	7

After mutation

1	2	3	4	5	4	5	6	7

What name is given to this type of chromosome mutation?

A inversion
B insertion
C translocation
D duplication

13. The following steps are involved in the process of genetic engineering.

1 insertion of a plasmid into a bacterial host cell
2 use of an enzyme to cut out a piece of chromosome containing a desired gene
3 insertion of a desired gene into a bacterial plasmid
4 use of an enzyme to open a bacterial plasmid

Which is the correct sequence of these steps?

A 2 4 3 1
B 4 1 2 3
C 4 2 1 3
D 2 3 4 1

14. Which of the following adaptations allows a plant to tolerate grazing by herbivores?

A Thick waxy cuticles on leaves
B Leaves reduced to spines
C Meristems low on the plant
D Stem with thorns

15. The graph shows the carbon dioxide exchange by the leaves of two plant species at different light intensities.

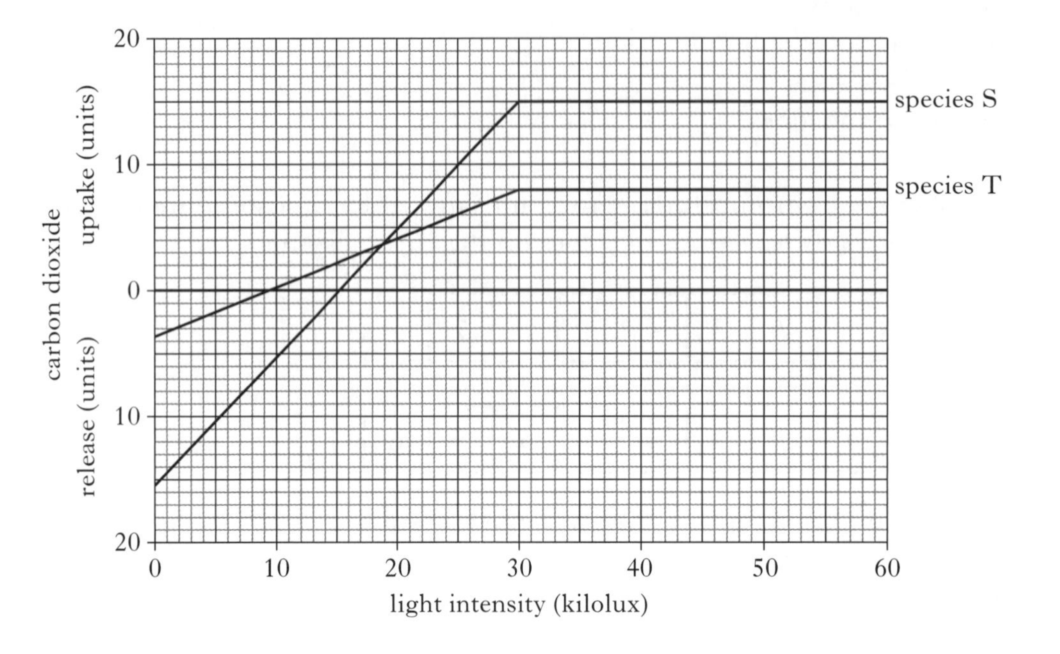

By how many kilolux is the compensation point for species S greater than the compensation point for species T?

A 4
B 5
C 7
D 11

16. In holly trees, the average number of spines per leaf decreases as the distance from the ground at which they grow increases.

The table below shows the results of an investigation in which the number of spines on a sample of leaves growing at different distances from the ground were counted.

Distance from ground	*Numbers of spines on a sample of leaves*
P	9, 13, 12, 10, 11
Q	10, 8, 13, 15, 14
R	10, 12, 9, 9, 10
S	9, 13, 15, 15, 13

Which of the following shows the distances from the ground in the correct order from the top of the tree to the bottom?

A P Q R S
B R P S Q
C S Q P R
D R P Q S

17. Peregrine falcons are predators which attack flocks of starlings. The graph shows how percentage attack success by peregrines varies with the number of starlings in a flock.

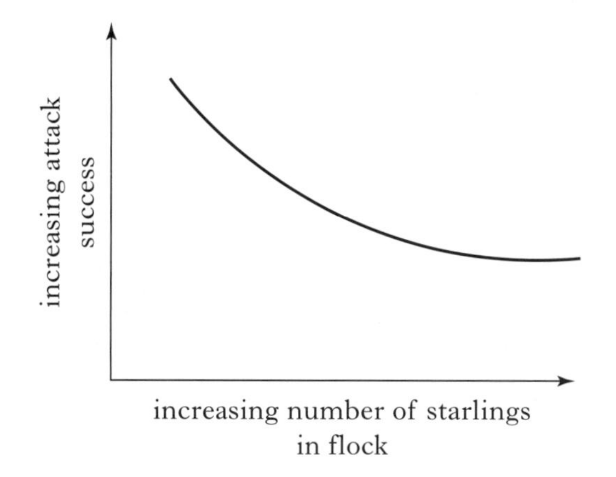

Which of the following statements could best explain the results shown on the graph?

A peregrine falcon

A finds hunting easier if a flock is larger
B prefers hunting a medium sized flock
C can hunt more selectively when a flock is larger
D is more effective in hunting if a flock is smaller.

18. The table shows the percentage recombination values for genes W, X, Y and Z on a chromosome.

Genes	*Percentage recombination*
W and Y	6
X and Y	14
X and Z	10
W and Z	30

Which is the correct sequence of these genes on the chromosome?

A Y W Z X
B Y W X Z
C W Y X Z
D W X Y Z

19. Which of the following show **two** adaptations of the leaves of xerophytic plants?

A Small surface area and thin waxy cuticles.
B Rolled and hairy.
C Large surface area and thick waxy cuticles.
D Hairy and many stomata.

20. Somatic fusion is a technique which is used to

A fuse cells from different plant species
B increase sexual incompatibility in plants
C fuse genetic material from a bacterium and a plant
D increase variability within a plant species.

21. The graph shows the increase in length of an organism over time.

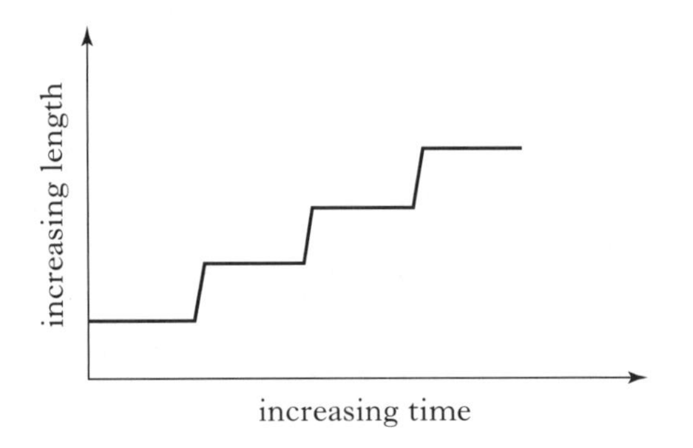

The best explanation of the appearance of the graph is that it shows

A seasonal growth of a perennial plant
B growth of an insect with several moults
C annual plant growth with several rapid growth periods
D human growth with several growth spurts.

22. The diagram below outlines part of the process which affects the control of metabolic rate in humans.

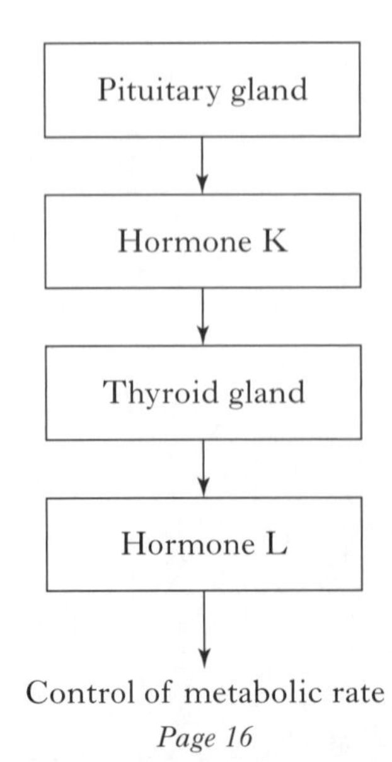

Which line in the table correctly identifies hormones K and L?

	Hormone K	*Hormone L*
A	Thyroid stimulating hormone (TSH)	Thyroxine
B	Growth hormone	Thyroxine
C	Thyroxine	Thyroid stimulating hormone (TSH)
D	Thyroid stimulating hormone (TSH)	Growth hormone

23. Increasing photoperiod affects the development of testes and ovaries and the onset of breeding behaviour in birds.

The graph below shows how the mass of the ovaries of a collared dove changed over 40 days from the start of its breeding period.

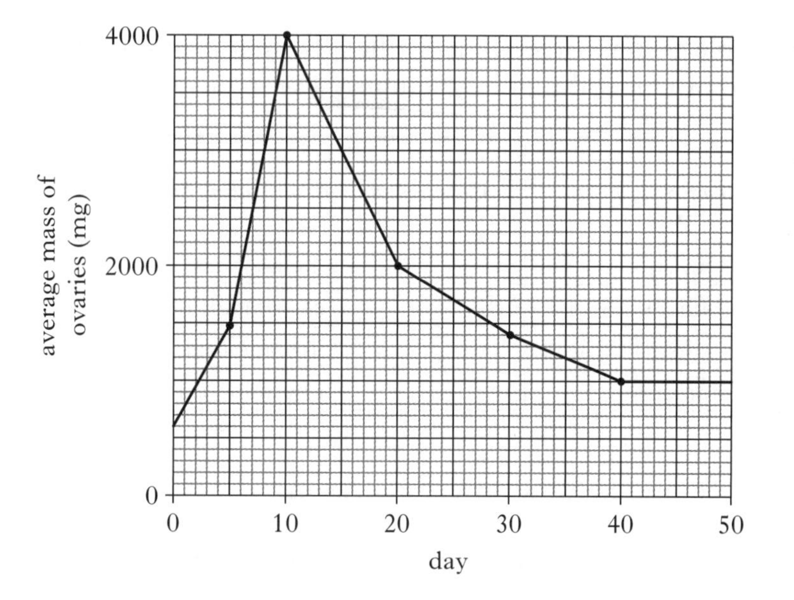

What was the percentage increase in the mass of the ovaries between Day 5 and Day 15?

A 25%
B 60%
C 100%
D 200%

24. The diagram below shows how temperature change affects the human skin.

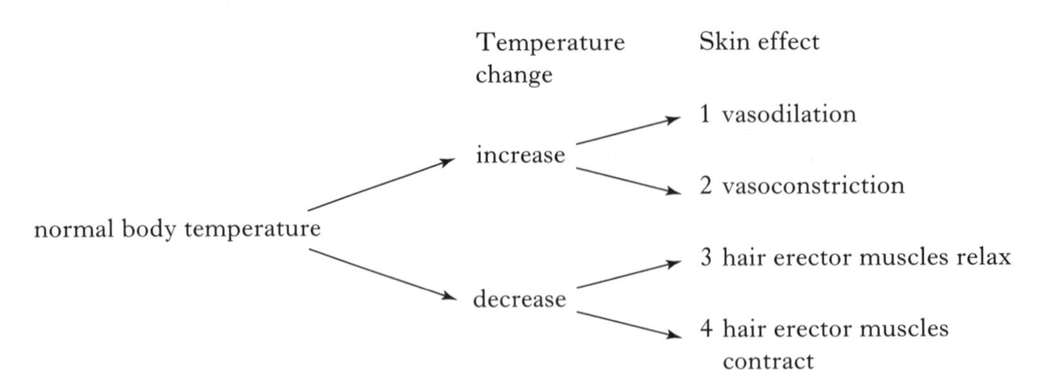

Which effects are likely to restore body temperatures to normal?

A 1 and 3
B 1 and 4
C 2 and 3
D 2 and 4

25. Which line in the table below correctly identifies the main source of body heat and the main method of controlling body temperature in an ectotherm?

	Main source of body heat	*Main method of controlling body temperature*
A	environment	behavioural
B	metabolism	physiological
C	metabolism	behavioural
D	environment	physiological

26. Populations may be monitored for a variety of reasons.

The list shows reasons for collecting biological data about plants.

1 Measurement of the biomass of a climax community
2 Estimation of the diversity of species in habitat
3 Conservation of an endangered species
4 Indication of the level of pollution in an ecosystem

For which of the reasons might a wild population of plants be monitored?

A 1 and 2
B 1 and 4
C 2 and 3
D 3 and 4

27. The graph below shows information about the numbers of grey and red squirrels in a woodland area over a 60 year period.

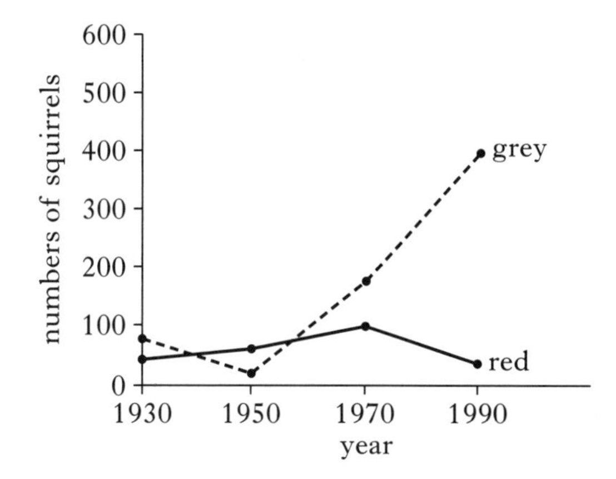

Which of the following statements about the squirrel populations is correct?

A The grey squirrels have always outcompeted red squirrels.
B The grey squirrel population steadily increased.
C The red squirrel population remained the same through the period.
D The red squirrel population increased during the 1940s.

28. The drawing shows part of an experiment set up to investigate mineral deficiency in seedlings.

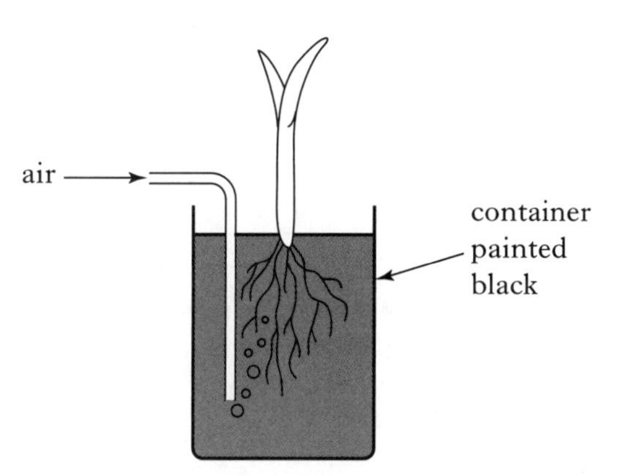

Which line in the table below correctly shows the reasons for the inclusion of the air bubbler and the black cover in the design of the experiment?

	Air bubbler	*Black cover*
A	encourages mixing of minerals	avoids algal growth in liquid medium
B	provides oxygen for respiration	avoids algal growth in liquid medium
C	encourages mixing of minerals	prevents photosynthesis in roots
D	provides oxygen for respiration	prevents photosynthesis in roots

29. The diagram below shows a metabolic pathway from a species of bacteria.

Substance P $\xrightarrow{\text{enzyme 1}}$ Substance Q $\xrightarrow{\text{enzyme 2}}$ Substance R $\xrightarrow{\text{enzyme 3}}$ Substance S

During an investigation of this metabolic pathway, an inhibitor of enzyme 2 was used.

Which of the graphs below shows the expected concentrations of substances Q and R when the inhibitor was added to a culture of the bacteria at time Y?

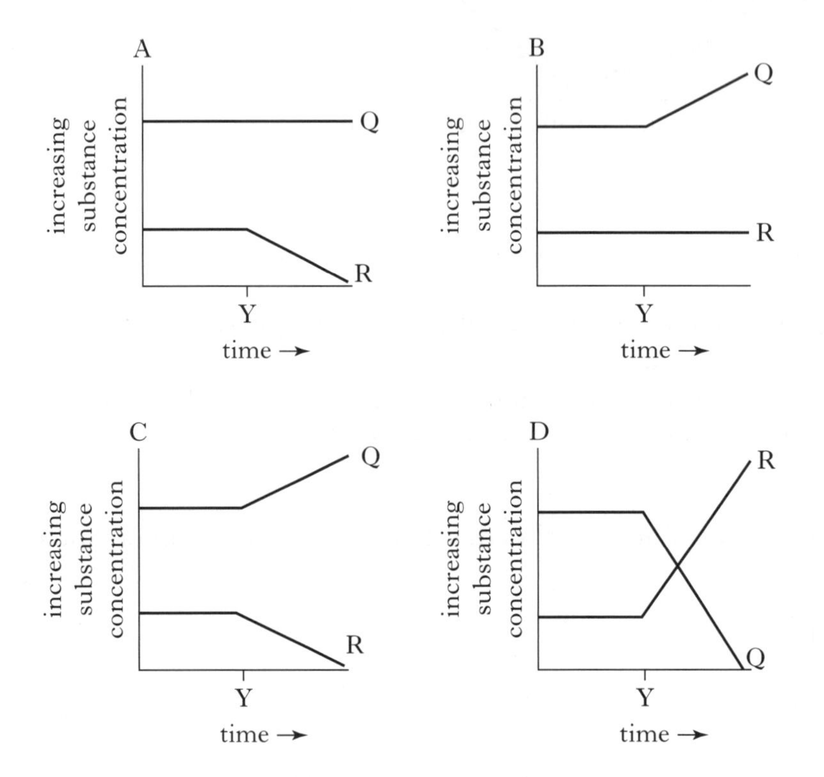

30. The diagram shows a section through a woody stem.

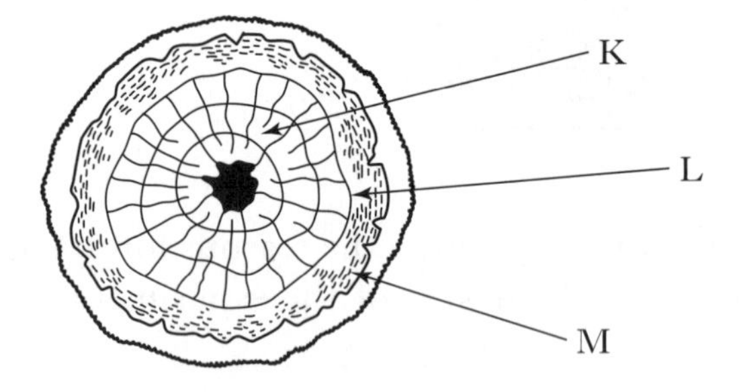

Which line in the table below correctly identifies the tissues K, L and M in the diagram?

	Tissue K	*Tissue L*	*Tissue M*
A	cambium	xylem	phloem
B	xylem	phloem	cambium
C	cambium	phloem	xylem
D	xylem	cambium	phloem

SECTION B

All questions in this section should be attempted.

All answers must be written clearly and legibly in ink in the spaces provided.

1. In an investigation, a piece of potato tissue was weighed then placed in a sucrose solution for five hours.

(*a*) The graph shows the change in the mass of the piece of tissue over the first four hours.

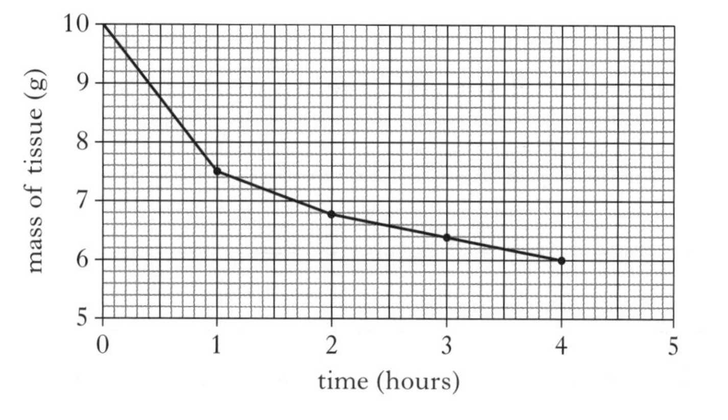

(i) Calculate the percentage loss in mass of the tissue between 1 and 4 hours after immersion.

Space for calculation

________________ % **1**

(ii) Predict the final mass of the tissue after being left for the full 5 hours.

____ **1**

(*b*) The diagram shows a single cell from the tissue one hour after immersion.

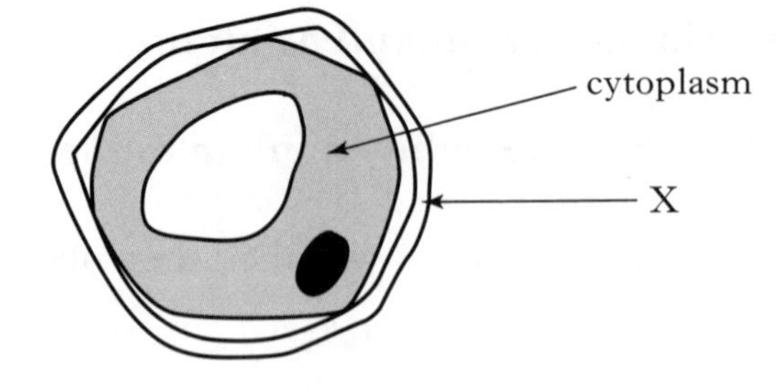

(i) Give the term used to describe the state of this cell.

________________________________ **1**

(ii) Explain the appearance of the cell in terms of the process of osmosis.

____

____

________________________________ **2**

(iii) Name the substance of which part X in the diagram is composed.

________________________________ **1**

2. The diagram shows the relationship between aerobic respiration and muscle contraction in mammals.

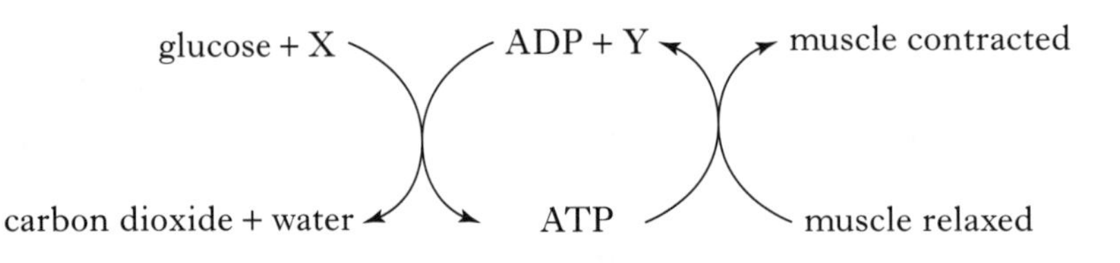

(*a*) (i) Name substances X and Y.

X________________

Y________________ **1**

(ii) Describe the role of ATP in the relationship shown in the diagram.

____

________________________________ **1**

(*b*) The drawing shows a mitochondrion.

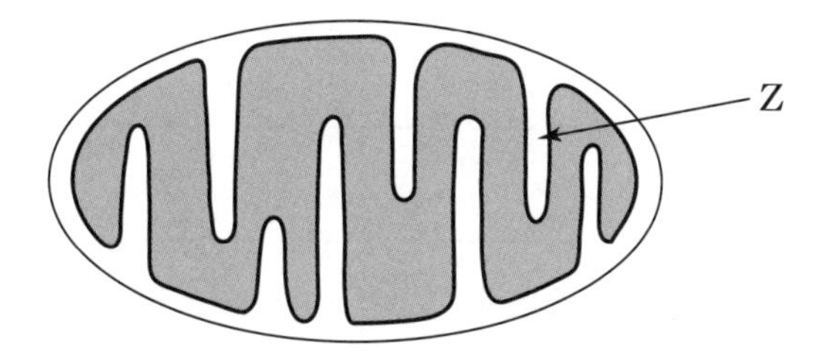

(i) Name part Z.

______________________________ **1**

(ii) Name the part of the mitochondrion in which the Krebs Cycle occurs.

______________________________ **1**

3. An investigation into the uptake of sodium ions by a unicellular organism was carried out.

Graph 1 shows the rate of uptake of sodium ions at 30°C by the unicellular organisms kept in solutions with different oxygen concentrations.

Graph 2 shows the rate of sodium ion uptake in constant oxygen concentration by the unicellular organisms kept in solutions at different temperatures.

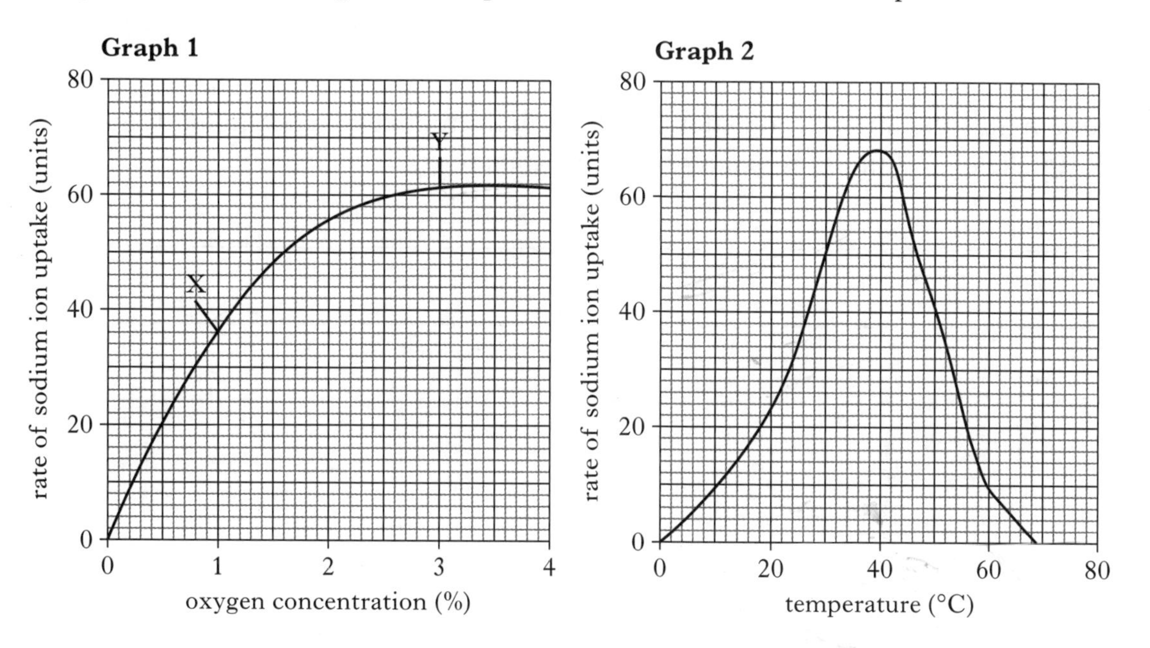

(*a*) (i) Use values from **Graph 1** to describe the changes in the uptake of sodium ions as the concentration of oxygen increased from 0 to 4%.

____

______________________________ **2**

(ii) How many times greater is the rate of sodium ion uptake at 2.5% oxygen than 0.5% oxygen?

Space for calculation

______________ % **1**

(*b*) (i) Give evidence from **Graph 1** which suggests that oxygen concentration is limiting the rate of sodium uptake at point X.

____

____ **1**

(ii) Suggest a reason why **Graph 1** levels off at point Y.

____

____ **1**

(*c*) Use information from **Graphs 2 and 1** to find the oxygen concentration at which the results in **Graph 2** were obtained.

______________% **1**

(*d*) Express as the simplest whole number ratio, the units of sodium ion uptake per minute at 10°C and 30°C.

Space for calculation

__________ : __________

at 10°C at 30°C **1**

(*e*) Explain how the information shown in **Graph 2** supports the statement that the uptake of sodium ions occurs by active transport.

____

____

____ **2**

4. The diagram shows detail of a piece of nucleic acid undergoing a cellular process.

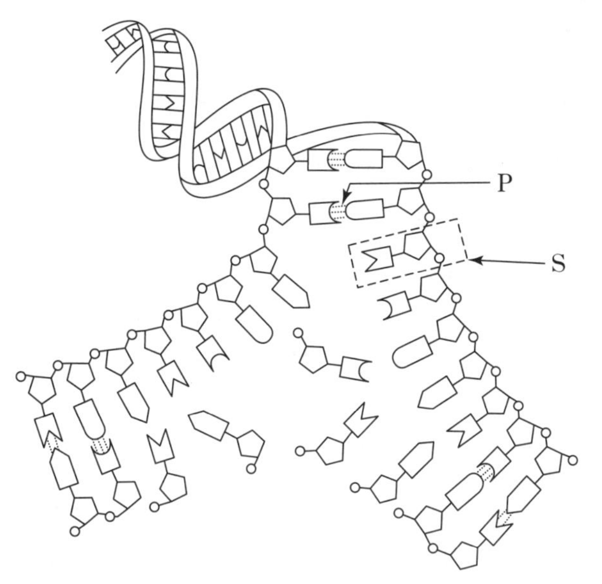

(*a*) (i) Name the structural unit of nucleic acid enclosed by the box S.

____ **1**

(ii) What type of bond is shown at P?

____ **1**

(*b*) A length of DNA contained 1200 bases of which 20% were cytosine.

Calculate how many adenine bases the length contained.

Space for calculation

____________ bases **1**

(*c*) Name the cellular process which is taking place in the diagram and justify your answer with **one** piece of evidence.

Process

____ **1**

Justification

____

____ **1**

5. In fruit flies, the allele determining grey body **G** is dominant to the allele determining black body **g** and the allele for normal wings **W** is dominant to the allele for vestigial wings **w**.

(*a*) Two flies both heterozygous for these genes were crossed.

Complete the grid below by adding the genotypes of:

(i) the male and female gametes; **1**

(ii) the other possible offspring. **1**

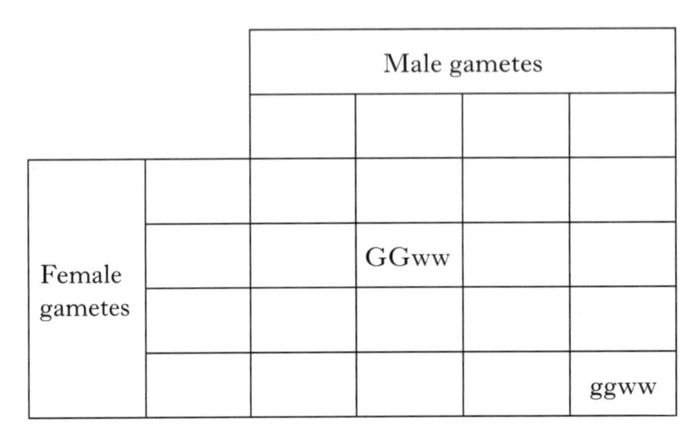

(*b*) Complete the blanks below to predict the phenotype ratios **expected** from this cross.

___ grey normal : ___ grey vestigial : ___ black normal : ___ black vestigial **1**

(*c*) After observing **actual** offspring from the cross, it was concluded that the genes involved were linked.

Describe how the **actual** phenotype ratio of the offspring could lead to this conclusion.

____

____ **2**

6. In Sickle Cell Disease, an abnormal form of the protein haemoglobin is produced because of a mutated allele of the haemoglobin gene.

The mutation causes the amino acid valine to be present at a specific point in the haemoglobin molecule instead of glutamic acid.

(*a*) Explain why a deletion of a DNA base could **not** have produced the effect described above.

____

____ **1**

(*b*) The table shows the DNA triplets for various amino acids.

Amino acid	*DNA Triplet*
Proline	GGC
Glutamic acid	CTC
Lysine	TTC
Threonine	TGC
Valine	CAC

The drawing shows the part of the amino acid sequence of haemoglobin that is affected by the mutated gene in Sickle Cell Disease.

Amino acid sequence

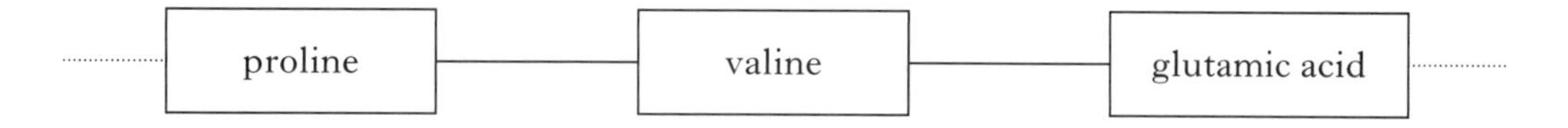

Give the sequence of bases on the three messenger RNA codons that would correspond to this amino acid sequence.

___ ___ ___   ___ ___ ___   ___ ___ ___ **1**

(*c*) Gene mutation is artificially increased in frequency when genetic material is exposed to mutagenic agents.

Name a mutagenic agent.

____ **1**

(*d*) The list shows processes which affect the gene pool of a species.

1 Mutation
2 Natural Selection
3 Isolation

(i) Use the numbers to complete the flow chart to show the correct sequence of these processes during the evolution of a new species.

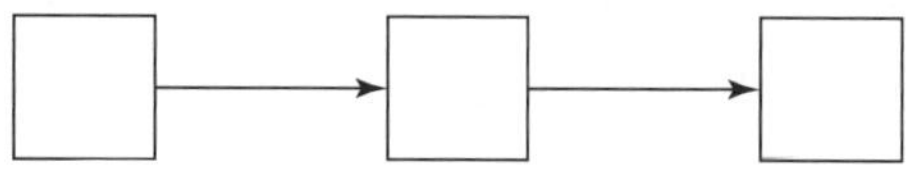

1

(ii) Apart from mutation, give another process which can affect variation within a species.

____ **1**

7. Many different species of duck occur in Scotland.

The diagram shows the heads of four species of duck and information about their feeding methods.

Mallard
Up-ends in shallow water to cut vegetation from the bottom.

Shoveler
Filters micro-organisms from shallow water.

Eider
Dives in sea water to scrape shellfish from rocks.

Goosander
Catches fish swimming under flowing water.

(*a*) (i) Explain how the information given about these ducks shows that they are adapted to avoid interspecific competition.

____

____ **1**

(ii) What further information would be needed about the four species of duck to conclude that they had evolved by adaptive radiation?

____

____ **1**

(*b*) The duck species have evolved in ecological isolation.

(i) Give two other types of isolating barrier involved in the evolution of new species.

1 ____

2 ____ **1**

(ii) Explain the importance of isolating barriers in the evolution of new species.

____

____ **1**

8. (*a*) The grid below shows adaptations of bony fish for osmoregulation.

A Active secretion of salts from cells in gills	B High glomerular filtration rate	C Many large glomeruli in kidney
D Low glomerular filtration rate	E Few small glomeruli in kidney	F Active uptake of salts by cells in gills

Use letters from the grid to answer the following questions.

(i) Which three adaptations would be found in fish living in hypertonic conditions?

Letters ________, ________, and ________ **1**

(ii) Which two adaptations would result in the production of high volumes of urine?

Letters ________ and ________ **1**

(*b*) The table shows adaptations of the desert rat connected with maintaining a water balance.

Tick (✓) the boxes which show behavioural adaptations designed to conserve water.

Adaptation	*Tick* (✓)
Lives in moist burrows	
Forages nocturnally	
Does not sweat	
Eats dry foods to gain some moisture	
Blood has high ADH levels	

2

9. Barn owls are territorial and feed on small mammal prey items.

They defend their territory using a variety of calls made from perches and while flying over the territory. Fights between birds sometimes occur.

(*a*) (i) Using the information given, explain the value to barn owls of defending a territory in terms of the economics of foraging behaviour.

____

____ **1**

(ii) Suggest one factor that could influence the size of a territory held by a pair of owls.

____

____ **1**

(*b*) The graph shows the percentages of the main prey items and the total monthly number of prey items in the diet of a pair of barn owls over a year.

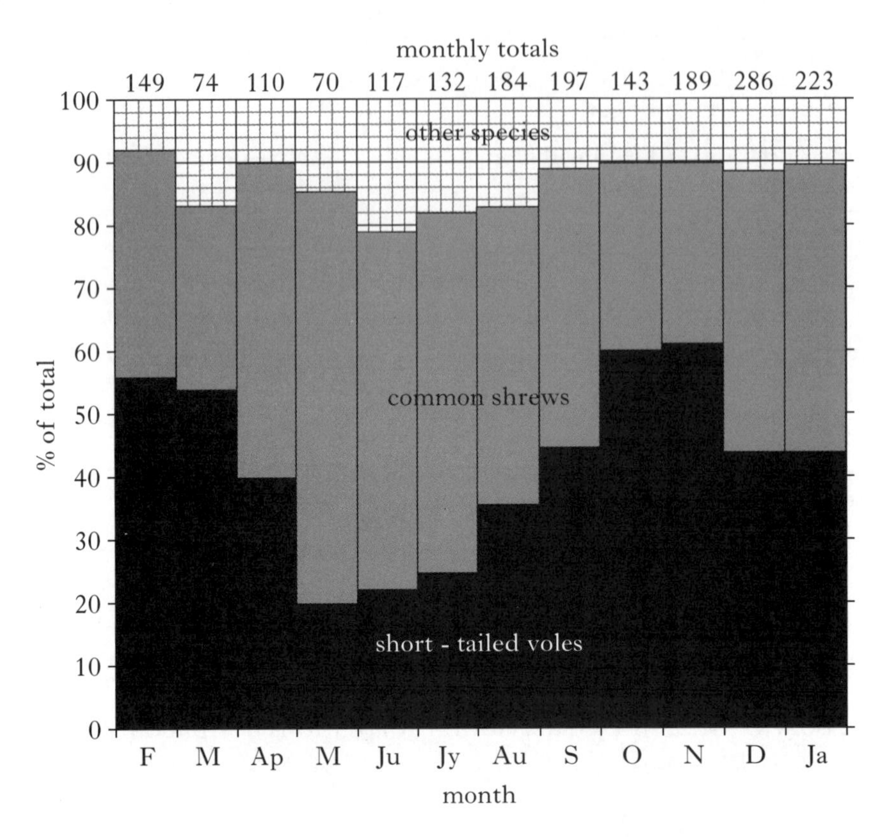

(i) Identify the month in which the common shrew is the commonest item in the diet of this owl pair.

______________________________ **1**

(ii) Calculate the number of common shrews in the diet in April.

Space for calculation

______________________________ **1**

(iii) Suggest why the numbers and proportions of different prey items varies throughout the year.

Numbers

____

______________________________ **1**

Proportions

____

______________________________ **1**

10. (*a*) The grid shows terms used in describing the control of lactose metabolism in *Escherichia coli*.

A Lactose digesting enzyme	B Structural gene	C Operator
D Regulator gene	E Repressor molecule	F Inducer

Use **a letter or letters** from the grid to answer the following questions.

(i) Which can combine with lactose?

____ **1**

(ii) Which controls synthesis of the repressor molecule?

____ **1**

(iii) In the absence of the inducer, which is **not** produced?

____ **1**

(*b*) Describe the part played by genes in metabolic pathways.

____

____

____ **2**

11. An experiment was carried out to investigate germination in wheat grains.

Wheat grains were soaked in water for 24 hours then washed in a solution to remove any bacteria and fungi on the surface of the grains. The grains were carefully cut and separated into parts containing the embryos and parts without embryos.

The parts were placed into four dishes containing equal volumes of distilled water either with or without GA. The dishes were kept in darkness at 20°C and, every 3 hours, the concentration of amylase in the solutions in the dishes was measured. The results are shown in the **Table**.

Table

Dish	*Grain Part*	*Distilled water*	*Amylase concentration* (units)				
			0 hours	3 hours	6 hours	9 hours	12 hours
A	without embryo	with GA	0.0	1.5	2.9	3.2	3.2
B	with embryo	with GA	0.0	1.4	2.7	3.1	3.2
C	without embryo	without GA	0.0	0.0	0.0	0.0	0.0
D	with embryo	without GA	0.0	0.5	1.2	1.5	1.8

(*a*) (i) Complete the table below to give a reason for each experimental procedure.

Experimental procedure	*Reason*
soak the grains in water for 24 hours	
wash the grains to remove bacteria or fungi	

2

(ii) Explain why some of the pieces were placed in dishes without GA.

____

______________________________ **1**

(iii) Identify one variable not already described which should be kept constant.

______________________________ **1**

(*b*) On the grid below, use values from the **Table** to plot a line graph by:

(i) adding a scale and label to each axis; **1**

(ii) presenting the results for dishes A and D **and** labelling the lines. **1**

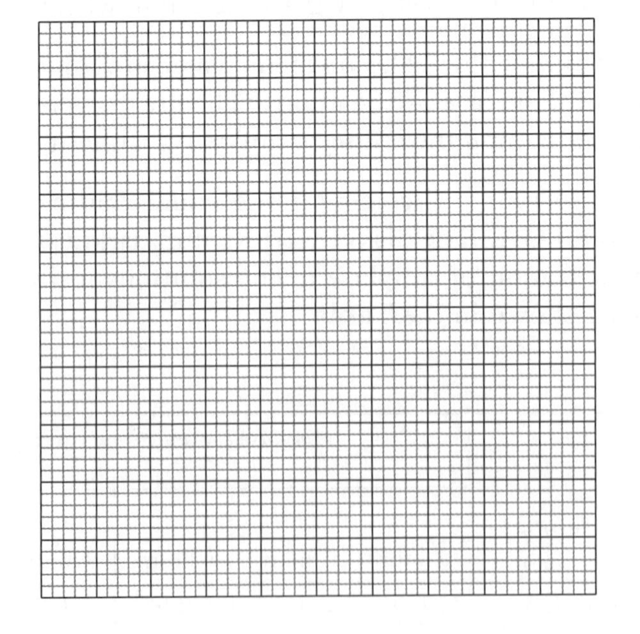

(*c*) Predict the concentration of amylase in dish D after 15 hours.

__________ units **1**

(*d*) From the results in dishes A and D, suggest a conclusion about the role of

(i) the embryo in germination of wheat grains.

____

____ **1**

(ii) GA in germination of wheat grains.

____

____ **1**

(*e*) Name the part of a wheat grain which produces amylase.

____ **1**

12. The table shows macroelements, their importance and symptoms of their deficiency in wheat plants.

(*a*) Complete the table by filling in the blank boxes. **3**

Macroelement	*Importance*	*Symptom(s) of deficiency*
phosphorus	found in DNA and ATP	
potassium		reduced growth
magnesium	found in chlorophyll	

(*b*) Light is important in the growth and development of plants.

Describe how light is involved in the following growth effects.

(i) Phototropism

____

____

____ **1**

(ii) Etiolation

____

____ **1**

13. Part of the homeostatic control of blood glucose concentration in humans is shown in the diagram.

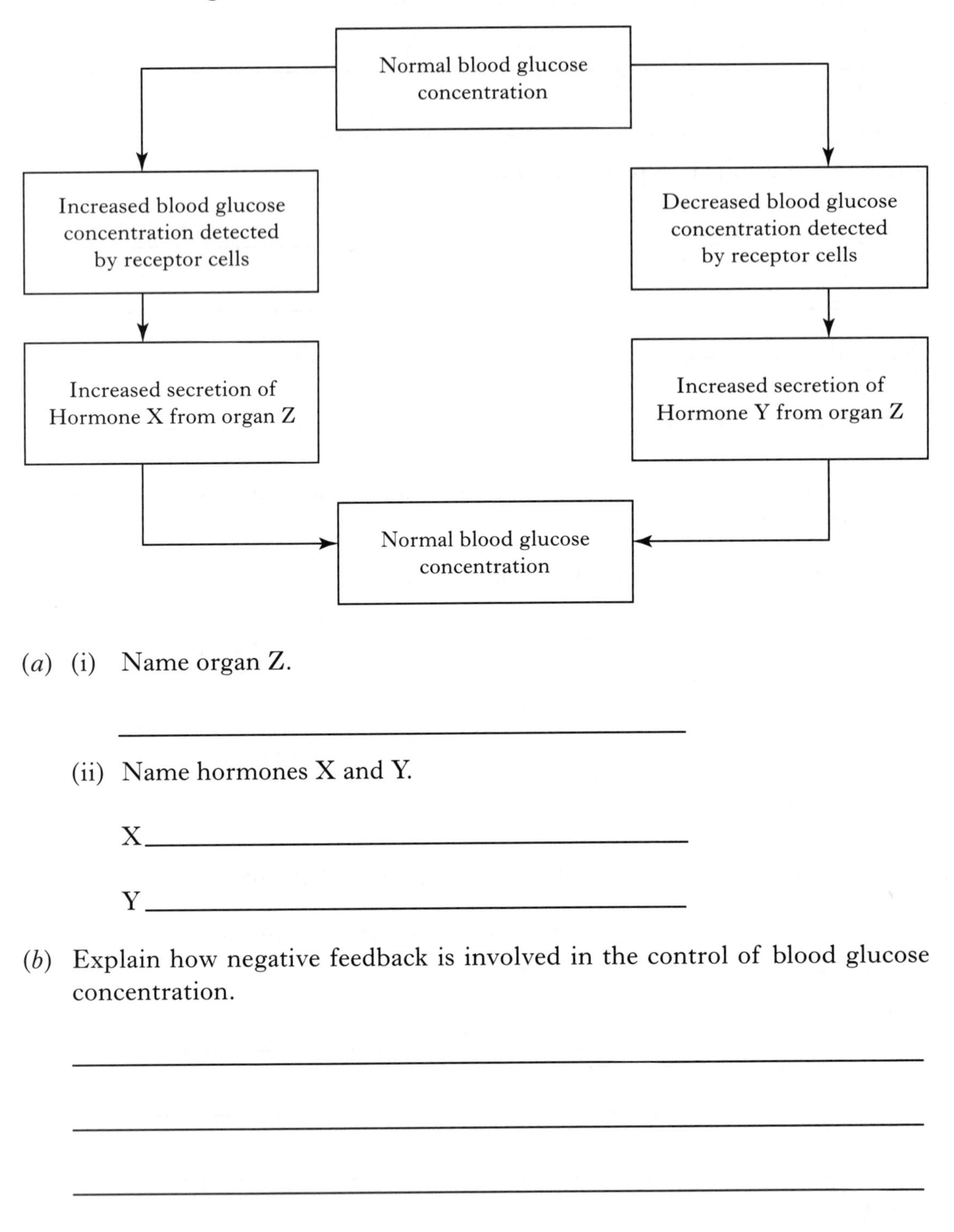

(*a*) (i) Name organ Z.

______________________________ **1**

(ii) Name hormones X and Y.

X______________________________

Y______________________________ **1**

(*b*) Explain how negative feedback is involved in the control of blood glucose concentration.

____

____

____ **2**

(*c*) The graph shows the changes in the blood glucose concentration in an individual with a normal pancreas and one with a damaged pancreas after each had ingested 100g of glucose. Normal blood glucose concentration is about 90 mg per 100 cm^3 blood.

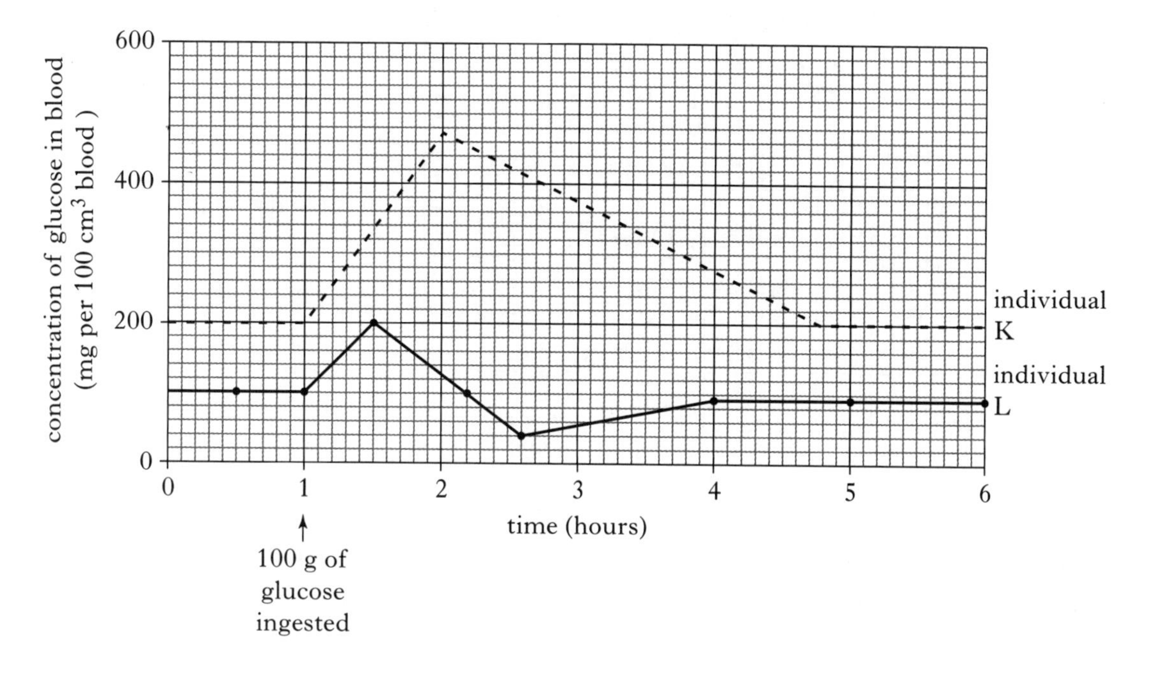

(i) Identify the individual with the damaged pancreas and justify your choice by using **two** pieces of evidence from the graph.

Individual ______________

Justification evidence

1

____ **1**

2

____ **1**

(ii) Calculate the average increase in glucose concentration per minute in the 60 minutes following ingestion of glucose by individual K.

Space for calculation

______________________ mg per 100 cm^3 blood per minute **1**

14. (*a*) The graph shows changes in the number of grey seal pups born on two British islands in the North Sea between 1950 and 2000.

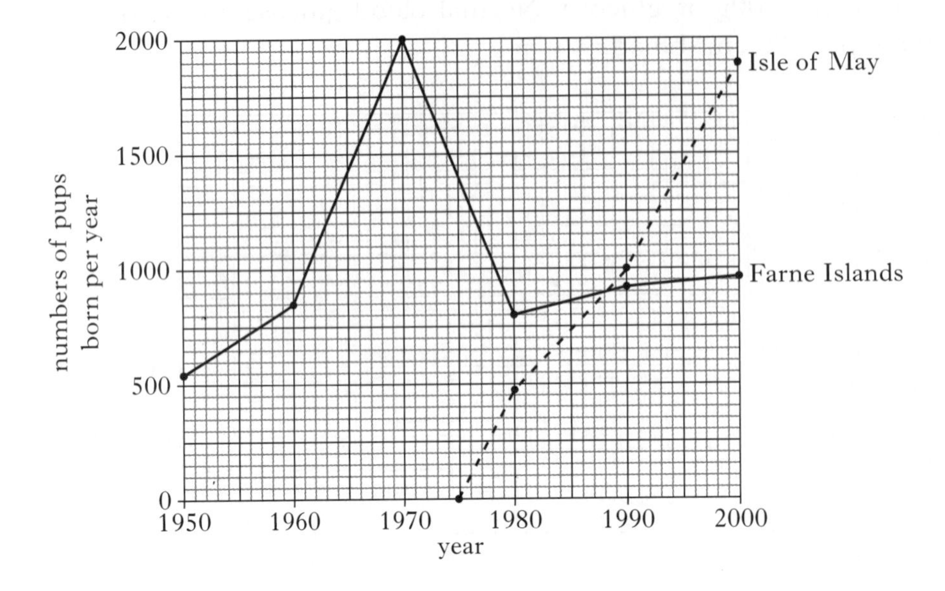

(i) Some seals were excluded from the Farne Islands after 1970 to prevent excessive soil erosion and these individuals then colonised the Isle of May.

What evidence is there that the exclusion from the Farne Islands increased the total population of grey seals in the North Sea?

____

____ **1**

(ii) Use figures from the graph to state the difference in grey seal births on the Farne Islands and the Isle of May in both 1980 and 1990.

1980

____

1990

____ **1**

(iii) During which 10 year period was there the greatest difference in seal numbers on either of the islands.

Tick the correct box.

1960 – 1970	1970 – 1980	1980 – 1990	1990 – 2000
☐	☐	☐	☐

1

(*b*) Grey seal populations can be affected by both density dependent and density independent factors.

Give an example of each of these factors.

(i) Density-dependent ____________________________________ **1**

(ii) Density-independent ___________________________________ **1**

SECTION C

Both questions in this section should be attempted.

Note that each question contains a choice.

All answers must be written clearly and legibly in ink on separate paper.
Labelled diagrams may be used where appropriate.

1. Answer **either** A **or** B.

A. Give an account of photosynthesis under the following headings:

(i) the absorption of light by photosynthetic pigments; **4**

(ii) the light dependent stage. **6**

(10)

OR

B. Give an account of the cellular response in defence in animals under the following headings:

(i) phagocytosis **5**

(ii) antibody production. **5**

(10)

In question 2, ONE mark is available for coherence and ONE mark is available for relevance.

2. Answer **either** A **or** B.

A. Give an account of the first and second stages of meiosis. **(10)**

OR

B. Give an account of the transpiration stream and its importance to plants. **(10)**

[End of Question Paper]

Exam 2

Biology Higher

Practice Papers
For SQA Exams

Exam 2

You have 2 hours, 30 minutes to complete this paper.

Try to answer all of the questions in the time allowed.

Write your answers in the spaces provided, including all of your working.

Leckie × Leckie
Scotland's leading educational publishers

SECTION A

All questions in this section should be attempted.

Answers should be given on the separate answer sheet provided on page 6

1. The diagram below shows the structure of a yeast cell highly magnified.

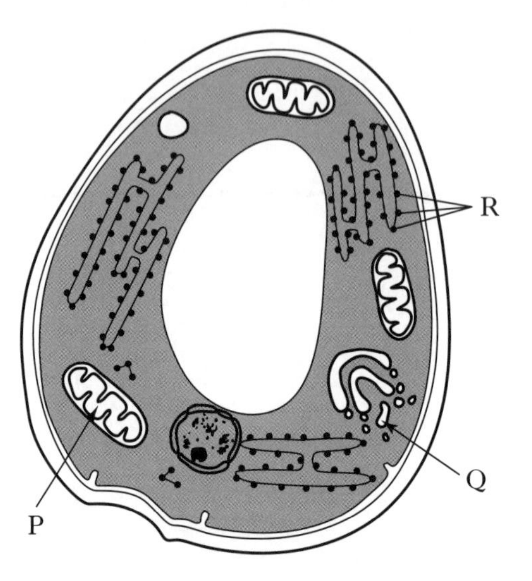

Which line of the table correctly matches each cell structure with its function?

	Aerobic respiration	*Protein synthesis*	*Packaging materials for secretion*
A	R	P	Q
B	P	R	Q
C	P	Q	R
D	Q	R	P

2. Visking tubing is selectively permeable.

The experiment shown below was set up to demonstrate osmosis.

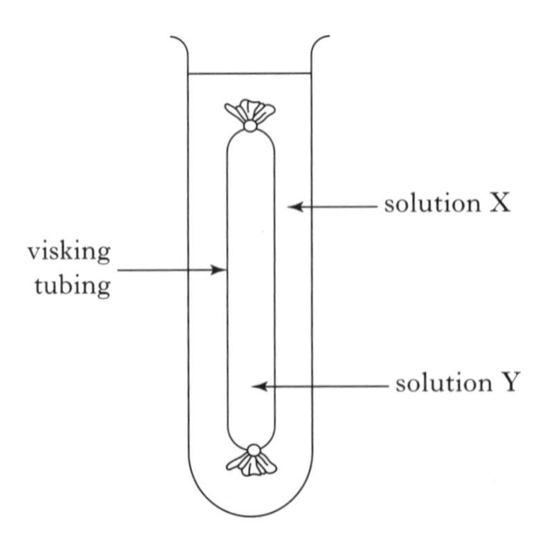

After 2 hours, the results shown in the table below were obtained.

Initial mass of Visking tubing + contents	9.0 g
Mass of Visking tubing + contents after 2 hours	7.8 g

The results shown could be obtained when

A Y is a 10% salt solution and X is a 5% salt solution.
B Y is a 5% salt solution and X is water.
C Y is a 5% salt solution and X is a 10% salt solution.
D Y is a 10% salt solution and X is water.

3. The diagram shows the fate of sunlight striking a leaf.

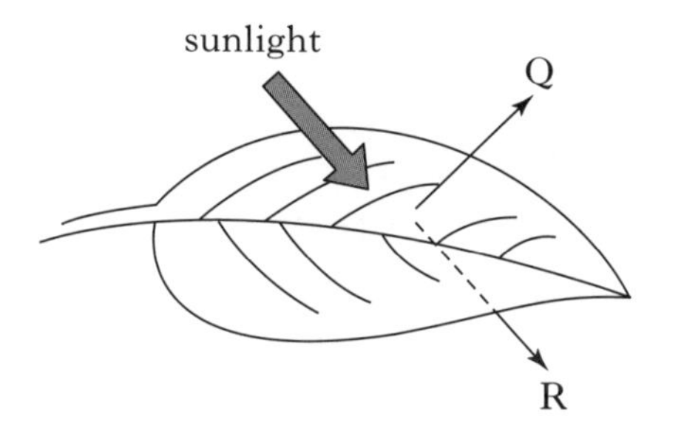

Which line in the table below identifies correctly the fates of sunlight shown by Q and R?

	Q	*R*
A	reflection	transmission
B	absorption	transmission
C	transmission	reflection
D	reflection	absorption

4. Photosynthetic pigments can be separated by means of chromatography and displayed as shown in the chromatogram below.

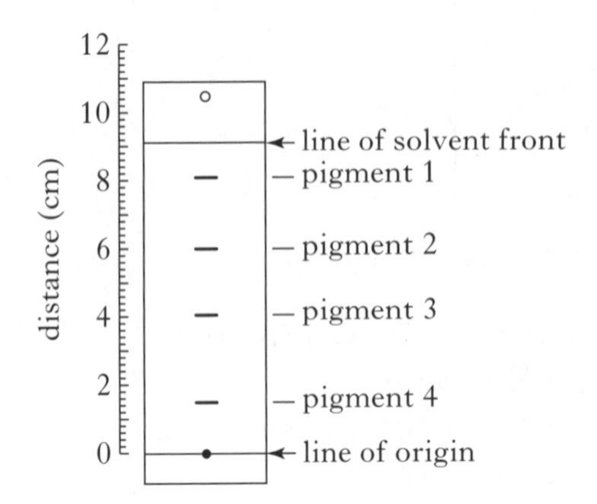

A pigment can be identified from its R_f value which can be calculated as follows:

$$R_f = \frac{\text{distance travelled by pigment from origin}}{\text{distance travelled by solvent from origin}}$$

Which line of the table correctly identifies the R_f values of pigments 1 and 2 on the above chromatogram?

	Pigment 1	*Pigment 2*
A	0.67	0.89
B	0.44	0.67
C	0.17	0.44
D	0.89	0.67

5. Which line in the table describes correctly **both** aerobic respiration and anaerobic respiration in plant cells?

	Aerobic Respiration	*Anaerobic Respiration*
A	There is a net gain of ATP	Carbon dioxide is not produced
B	There is a net gain of ATP	Oxygen is required
C	Carbon dioxide is produced	Ethanol and carbon dioxide are produced
D	Lactic acid is produced	Ethanol and carbon dioxide are produced

6. The diagram shows a mitochondrion.

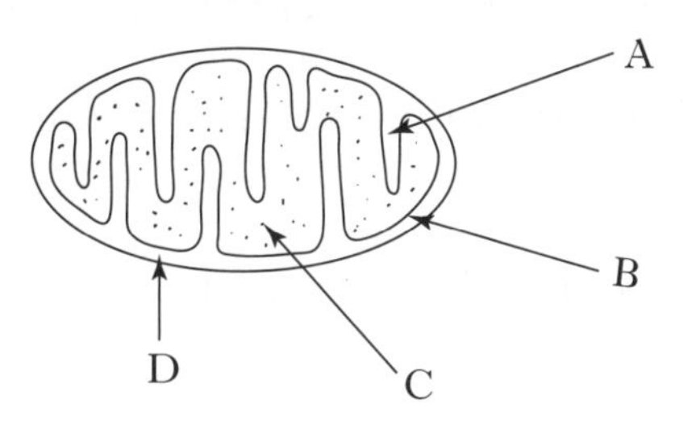

In which region does Krebs cycle take place?

7. Which of the following is a fibrous protein?

A Amylase
B Collagen
C Cellulose
D Insulin

8. DNA controls cell activities by coding for the production of

A proteins
B bases
C carbohydrates
D amino acids.

9. A fragment of DNA was found to have 30 guanine bases and 15 adenine bases.

What is the total number of deoxyribose sugar molecules in this fragment?

A 90
B 45
C 30
D 180

10. Which of the following is **not** a cellular response by plants to invasion by other organisms?

A Production of tannins
B Secretion of nicotine
C Secretion of resin
D Production of antibodies

11. A tall snapdragon plant with red flowers was crossed with a dwarf snapdragon plant with yellow flowers. The F_1 generation were all tall with red flowers.

An F_1 generation plant was then self pollinated and produced 320 offspring in the F_2 generation.

Which line in the table identifies correctly the most likely phenotype ratio in the F_2 generation.

	Tall with red flowers	*Tall with yellow flowers*	*Dwarf with red flowers*	*Dwarf with yellow flowers*
A	174	65	61	20
B	174	0	0	146
C	80	80	80	80
D	106	52	54	108

12. Chiasma formation

A allows the pairing of homologous chromosomes
B allows gene exchange between homologous chromosomes
C results in the halving of the chromosome number
D results in the independent assortment of chromosomes

13. In pea plants, the allele for tall stems T, is dominant to its allele for dwarf stems t and the allele for hairy stems H is dominant to its allele for smooth stems h.

The following cross was carried out.

TtHh × tthh

64 offspring were produced from this cross.

How many offspring would be expected to have tall, smooth stems?

A 48
B 32
C 16
D 8

14. The term linkage refers to genes which are

A present on the same chromosome
B transferred from one chromosome to its homologous partner
C transferred from one homologous pair to another
D present on different chromosomes

15. Recombination frequencies of linked genes provide information on

A the genotype for a particular characteristic
B the mutation rate of the genes
C whether genes are recessive or dominant
D the order and location of genes on a chromosome.

16. Which of the following is not a source of genetic variation in sexual reproduction?

A Crossing over
B Mutation
C Non-disjunction
D Mitosis

17. The base sequence of a short piece of DNA is shown below.

T C G A A T G C

During replication, an inversion mutation occurred on the complementary strand.

Which of the following shows the mutated complementary strand?

A A G C T T A C T

B T C G A A T C G

C A G C T T A G C

D A G C T T A C G

18. Speciation has taken place when a sub-population

A is isolated from the rest of the population by a barrier to gene exchange
B can no longer interbreed successfully with the rest of the population
C shows increased variation due to mutation.
D is subjected to increased selection pressures in its habitat.

19. The dark form of the peppered moth became more common in areas of Britain where there was an increase in atmospheric pollution due to the burning of coal during the industrial revolution.

The increase in the dark form was due to

A dark moths migrating to areas which gave the best camouflage
B a change in the prey species taken by birds
C an increase in the mutation rate
D a change in selection pressure.

20. The following factors affect the transpiration rate in plants.

1 decreasing wind speed
2 increasing humidity
3 increasing light intensity
4 increasing temperature

Which two of these factors would cause a decrease in transpiration rate?

A 1 and 2
B 1 and 3
C 2 and 4
D 3 and 4

21. Plants grown in the absence of light have

A green leaves and short internodes
B green leaves and long internodes
C yellow leaves and short internodes
D yellow leaves and long internodes.

22. A long day plant will flower only if the number of hours of

A darkness is less than a critical value
B light is less than 12 hours
C light is less than a critical value
D darkness is more than a critical value.

23. Plants require macro-elements for the synthesis of various substances.

Which line in the table below shows the macro-elements required for synthesis of the substances shown?

	Substances		
	Chlorophyll	*Nucleic acid*	*RuBP*
A	phosphorus	magnesium	nitrogen
B	magnesium	nitrogen	phosphorus
C	phosphorus	nitrogen	magnesium
D	magnesium	phosphorus	nitrogen

24. The average daily dietary iron requirement of an individual was 0.5 mg. This increased to 1.5 mg when the individual started training to sprint at altitude.

What percentage increase does this represent?

A 33%
B 50%
C 200%
D 300%

25. If the concentration of glucose in the blood of a healthy man or woman decreases below normal, their pancreas produces

A more insulin but less glucagon
B more insulin and more glucagon
C less insulin but more glucagon
D less insulin and less glucagon.

26. Which of the following would result from a decreased level of ADH in the bloodstream of a man?

A Production of urine with a higher concentration of urea
B An increase in the permeability of kidney tubules
C A decrease in rate of glomerular filtration
D An increase in volume of urine produced

27. The table gives information on the concentration of urea in glomerular filtrate and urine.

	Glomerular filtrate	*Urine*
Urea concentration ($g/100\ cm^3$)	0.03	2.1

How many times more concentrated is the urea in urine than in glomerular filtrate?

A 0.7
B 7
C 70
D 700

28. The table shows the results of soaking oat shoots and oat roots in various concentrations of IAA. The initial lengths of both shoots and roots was 10 mm.

Concentration of IAA in bathing solution (molar)	*Length after 2 days (mm)*	
	Shoots	Roots
0	12.2	12.3
10^{-10}	12.1	13.4
10^{-9}	12.5	12.5
10^{-8}	13.3	11.2
10^{-7}	14.0	10.7

Which of the following conclusions can be drawn from the data?

The 10^{-8} molar solution of IAA

A stimulates growth in both shoots and roots
B stimulates growth in shoots and inhibits growth in roots
C inhibits growth in shoots and stimulates growth in roots
D inhibits growth in both shoots and roots.

29. In an experimental plot of one species of grass, the number of plants and the number of seeds produced were counted each year for two years.

Year	*Plants per 0.25 m²*	*Seeds per 0.25 m²*
1	270	702
2	1250	2375

The data suggests that in Year 1 compared with Year 2, the number of seeds per plant

A was lower due to less competition
B was higher due to less competition
C was lower due to more competition
D was higher due to more competition

30. During succession a number of changes take place in a plant community.

Which line of the table below correctly describes some of these changes?

	Changes in plant community		
	Biomass	*Food web complexity*	*Species diversity*
A	decrease	increases	increases
B	increases	increase	increases
C	decreases	decrease	increases
D	increases	increases	decreases

SECTION B

All questions in this section should be attempted.
All answers must be written clearly and legibly in ink in the spaces provided.

1. (*a*) The diagrams below show a chloroplast and an outline of the carbon fixation stage of photosynthesis.

chloroplast **carbon fixation stage**

P

GP
hydrogen
compound Q
RuBP
glucose

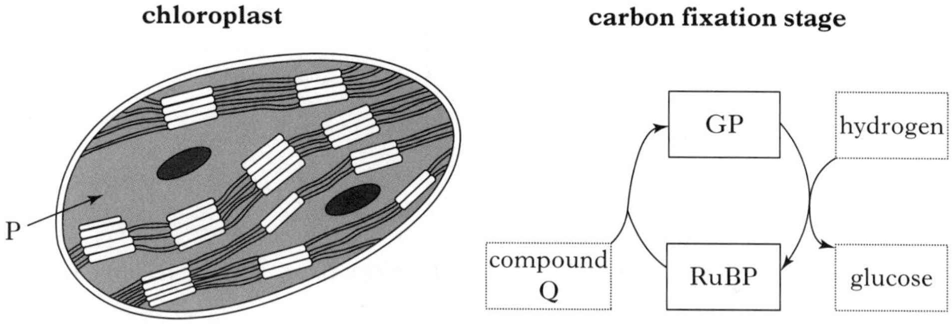

(i) Name region P and compound Q.

Region P ______________________________

Compound Q ______________________________ **1**

(ii) The hydrogen required for the carbon fixation stage comes from the light dependent stage of photosynthesis.

Give the source of the hydrogen.

______________________________ **1**

(*b*) Describe the role of ATP in photosynthesis.

____

____

______________________________ **1**

(*c*) Other than carbohydrates, name one major biological molecule produced as a result of photosynthesis.

______________________________ **1**

(*d*) The graph shows the relative concentrations of glycerate phosphate (GP) and ribulose 1,5-bisphosphate (RuBP) in a chloroplast as conditions change from a light period to a dark period and back to a light period.

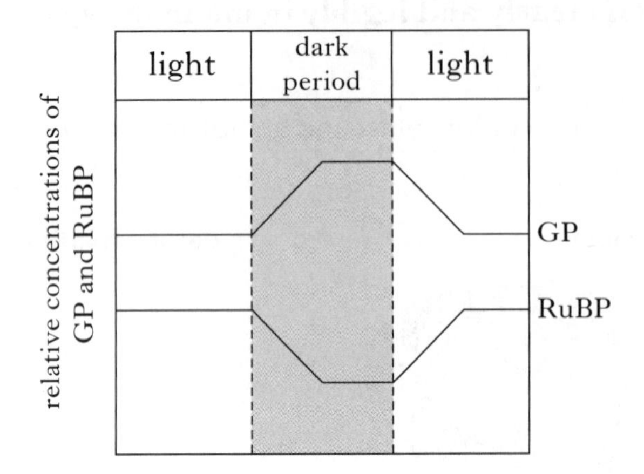

Explain the change in RuBP concentration over the first part of the dark period.

____

____

____ **2**

2. The diagram below represents an outline of the process of respiration in human muscle tissue.

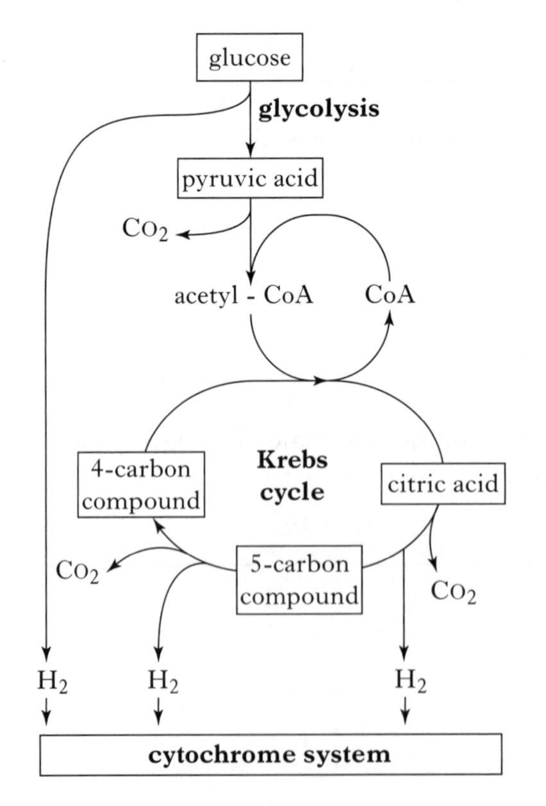

(*a*) Complete the table by giving the number of carbon atoms present in one molecule of each substance.

Substance	*Number of carbon atoms in one molecule*
Glucose	
Pyruvic acid	
Acetyl group	
Citric acid	

2

(*b*) Name a respiratory substrate, other than glucose, which can be used by muscle cells.

____ 1

(*c*) Apart from a suitable respiratory substrate and enzymes, what chemical substance is essential for glycolysis to occur?

____ 1

(*d*) Name the carrier that transfers hydrogen to the cytochrome system.

____ 1

(*e*) Describe the role of oxygen in aerobic respiration.

____

____

____ 1

3. The diagram shows how a short section of deoxyribonucleic acid (DNA) determines the sequence of amino acids in a protein molecule.

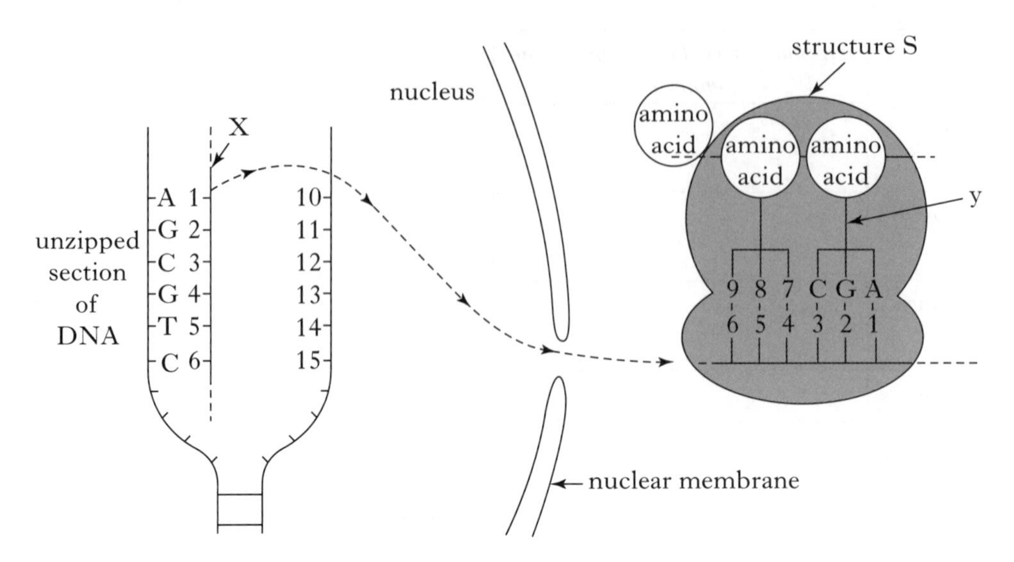

(*a*) Name molecules X and Y

X ______________________________

Y ______________________________ **1**

(*b*) Name structure S

______________________________ **1**

(*c*) Identify the bases 1, 8 and 15

1 ______________________________

8 ______________________________

15 ______________________________ **2**

(*d*) Describe the function of the Golgi apparatus in a cell.

____

______________________________ **1**

(*e*) The sequence of triplets on a strand of DNA coding for a polypeptide is shown below.

ATTTCACCGTACCAATAG

How many of the tRNA molecules involved in the synthesis of the polypeptide will have anti-codons with at least one uracil base?

______________________________ **1**

(*f*) If 28% of the bases in a particular sample of DNA are adenine, what percentage of its bases will be cytosine?

Space for calculation

______________________ % **1**

4. The diagram shows stages that occur when a cell is invaded by a virus.

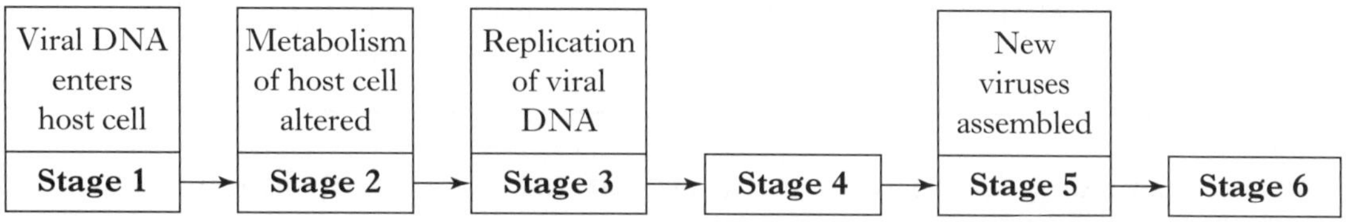

Viral DNA enters host cell	Metabolism of host cell altered	Replication of viral DNA		New viruses assembled	
Stage 1 →	**Stage 2** →	**Stage 3** →	**Stage 4** →	**Stage 5** →	**Stage 6**

(a) Describe the processes which take place during Stages 4 and 6.

Stage 4 ____

____

____

Stage 6 ____

____

____ **2**

(*b*) Name **one** substance which is supplied by the host cell during stage 3.

____ **1**

(*c*) Information about the reproduction and transmission success of a specific plant virus is given below.

- A plant cell infected by one virus always releases 100 new viruses
- 20% of the new viruses released go on to infect new plant cells.

Calculate how many viruses would eventually be released from these newly infected cells.

Space for calculation

____ **1**

(*d*) The grid shows substances involved in the cellular defence of organisms.

A resin	B antibody	C lysosome enzymes
D tannin	E antigen	F nicotine

Use letters from the list to identify a substance produced

(i) by plants which acts as barrier to prevent the spread of infection.

______________________________ **1**

(ii) following a transplant operation which can lead to rejection.

______________________________ **1**

(*e*) What treatment is given to reduce the risk of tissue rejection following a transplant operation?

______________________________ **1**

(*f*) The graph shows the primary and secondary immune responses triggered following infections by the same viruses.

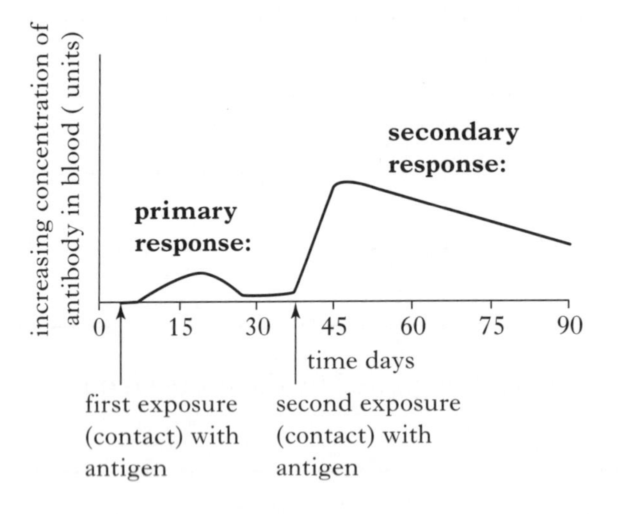

From the graph, describe **three** differences between the primary and secondary immune responses.

1 ______________________________

2 ______________________________

3 ______________________________ **2**

5. In Drosophila, the sex chromosomes in males are X and Y and in females XX. The gene for eye colour is sex-linked and the red-eyed allele is dominant to the white-eyed allele.

Crosses between two different sets of parents were carried out. The numbers and phenotypes of the offspring obtained are shown in the table below.

Cross	*Phenotypes of parents*	*Numbers and phenotypes of offspring*			
		Red-eyed males	*Red-eyed females*	*White-eyed males*	*White-eyed females*
1	White-eyed male × Red-eyed female	23	19	18	18
2	Red-eyed male × White-eyed female	0	35	34	0

(*a*) Complete the table below to show the genotypes of the parents in each of the above crosses.

Space for working

Cross	*Male genotype*	*Female genotype*
1		
2		

2

(*b*) Identify the cross, shown in the table, that allows the sex of the offspring to be determined by their eye-colour and justify your answer.

Cross ________________________________

Justification

____

____

____

1

(*c*) Describe where sex-linked genes are located.

1

(*d*) Explain why sex-linked conditions such as red green colour deficiency in humans are less common in females than in males.

1

6. Radish and cabbage plants both have a diploid chromosome number of 18. Crossing radish with cabbage produces infertile hybrids as shown.

parents	cabbage	radish
diploid chromosome number of parent plants	18	18
chromosome number of gametes	9	9
chromosome number of infertile hybrid	18	

(*a*) Explain why the hybrid plants are unable to produce gametes.

1

(*b*) An infertile hybrid can undergo a chromosome mutation which results in polyploidy.

Name a chromosome mutation which can result in polyploidy.

1

(*c*) Give **one** advantage for humans of polyploidy in crop plants.

1

7. The following statements refer to the Hawaiian Islands and a group of bird species called sicklebills which have evolved there.

Statements

1 *The Hawaiian Islands are extremely isolated, being 3500 km from the nearest mainland.*

2 *The Hawaiian sicklebills are descended from a finch-like ancestor.*

3 *Some sicklebill species have retained finch-like beaks and eat seeds, others have long thin beaks to reach nectar and others have intermediate beaks.*

(*a*) Name the type of isolation mechanism illustrated in Statement **1**.

____ **1**

(*b*) Explain how the information above illustrates adaptive radiation.

____

____

____

____ **2**

(*c*) Describe the advantage that might be gained by the sicklebills with intermediate beaks.

____

____ **1**

8. Some of the stages of a genetic engineering procedure to insert a specific gene from a donor organism into a bacterial cell are shown below.

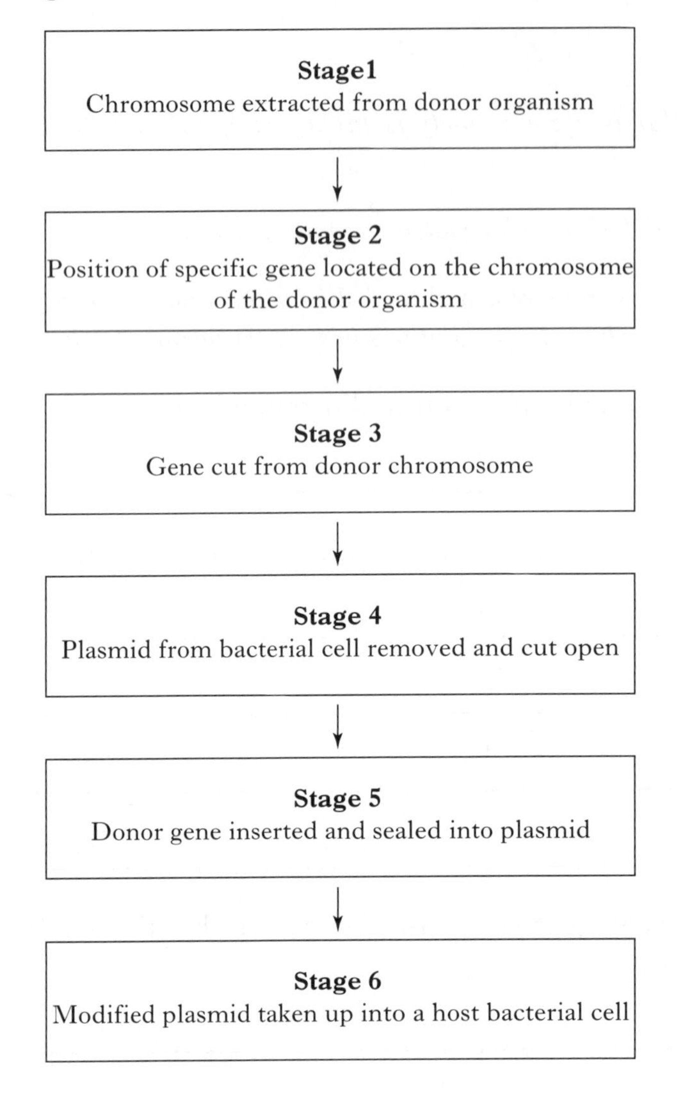

(*a*) Name a technique which can be used to locate the position of a specific gene on a chromosome.

____ **1**

(*b*) Complete the table by naming the enzymes involved in Stages 3, 4 and 5.

Stage	*Name of enzyme*
3	
4	
5	

2

9. The graph represents the growth pattern of an annual plant divided into periods.

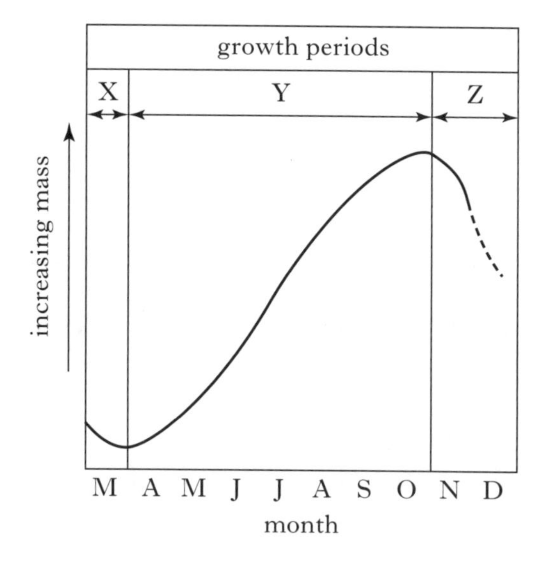

(*a*) Explain the changes in growth shown in periods X, Y and Z.

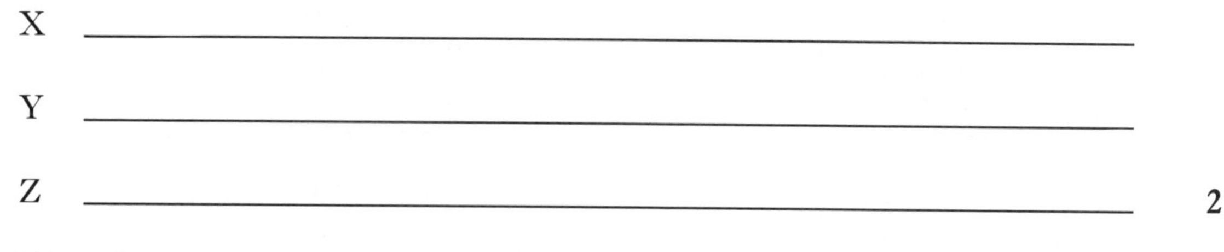

X ____

Y ____

Z ____ **2**

(*b*) The diagram shows a section through a woody stem with a small area magnified.

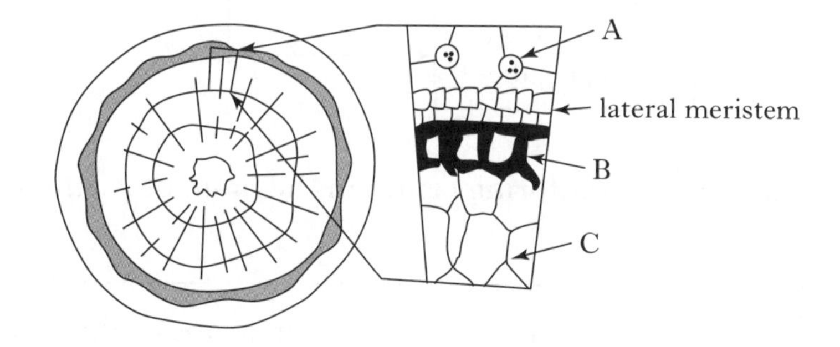

(i) Name the lateral meristem shown in the diagram and describe its function.

Name ____ **1**

Function ______________________________________

____ **1**

(ii) Identify the letter in the magnified area which indicates a xylem vessel formed in spring.

Give a reason for your answer.

Letter ______________________

Reason ______________________

______________________ 1

10. (*a*) The flow chart represents homeostatic control of body temperature in a mammal.

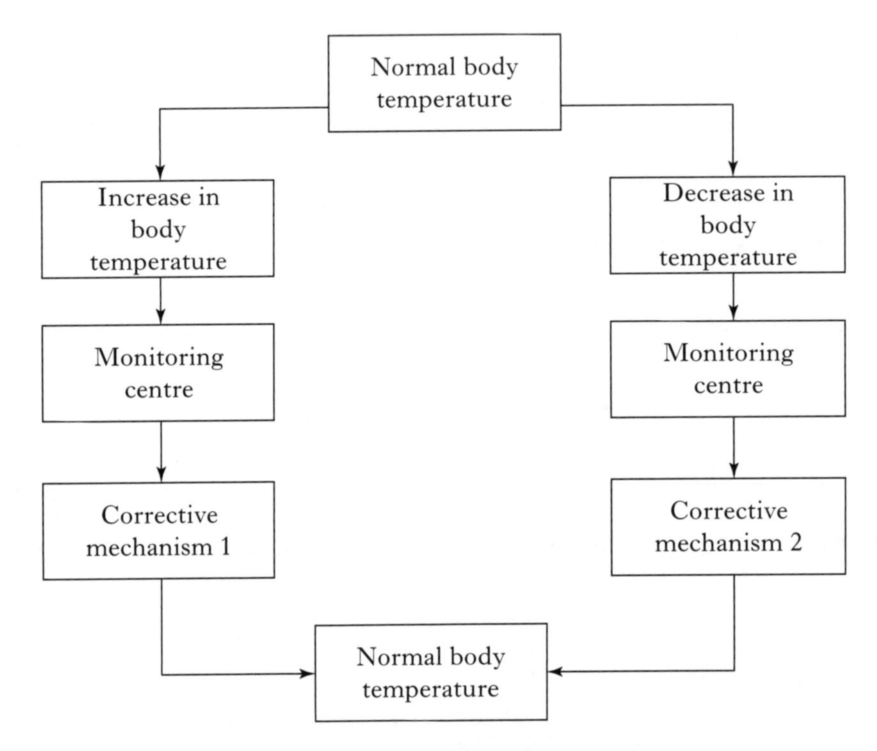

(i) Name the temperature monitoring centre in the brain of mammals.

______________________ 1

(ii) Underline one of the alternatives in each pair to make the following sentence correct.

In **corrective mechanism** 2 given in the flow chart, {vasoconstriction / vasodilation} results in {increased / decreased} blood flow to the skin. 1

(*b*) Explain how negative feedback is involved in the homeostatic control of body temperature.

____

____

____ **2**

(*c*) What term is used for animals that obtain most of their body heat from their own metabolism?

____ **1**

(*d*) Explain the importance of maintaining body temperature to human metabolism.

____

____ **1**

11. (*a*) The graphs provide information on the flock size of wading birds in relation to attack success by peregrine falcons.

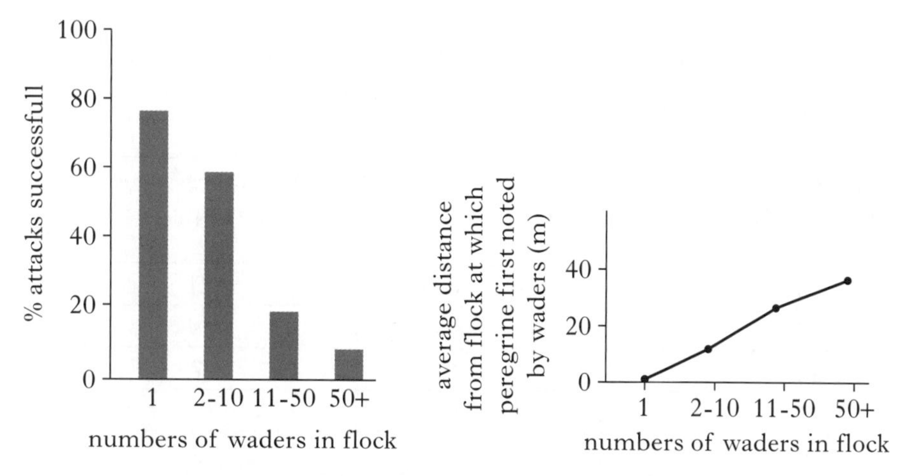

(i) State the relationship between the flock size and attack success and use the information in the graphs to explain the results.

Relationship ____

____ **1**

Explanation ____

____ **1**

(ii) Suggest a benefit, not shown in the graph, which wading birds gain by feeding together in a large flock rather than feeding alone.

____

____ **1**

(*b*) When first exposed to a harmless stimulus, a group of animals responded by showing avoidance behaviour.

When the stimulus was repeated several times the response to the harmless stimulus decreased.

Name this type of learned behaviour and state its advantage to the animals?

Name ____

Advantage ____

____ **1**

12. An investigation was carried out into the effect of different concentrations of salt solution on the mass of carrot tissue.

Six cubes of carrot were dried and weighed then each placed into salt solutions for two hours.

The cubes were reweighed and the percentage change in mass was calculated.

The results are shown in the table.

Carrot cube	*Concentration of salt solution (mol dm^{-3})*	*Initial mass (g)*	*Final mass (g)*	*Percentage change in mass (%)*
A	0.10	5.70	6.90	+21
B	0.15	6.50	7.41	+14
C	0.20	5.90	6.31	+7
D	0.25	6.30	6.30	0
E	0.30	5.90	5.43	−8
F	0.40	6.40	5.06	−21

(*a*) On the grid, plot a line graph to show the percentage change in mass of the carrot cubes against concentration of salt solution. **2**

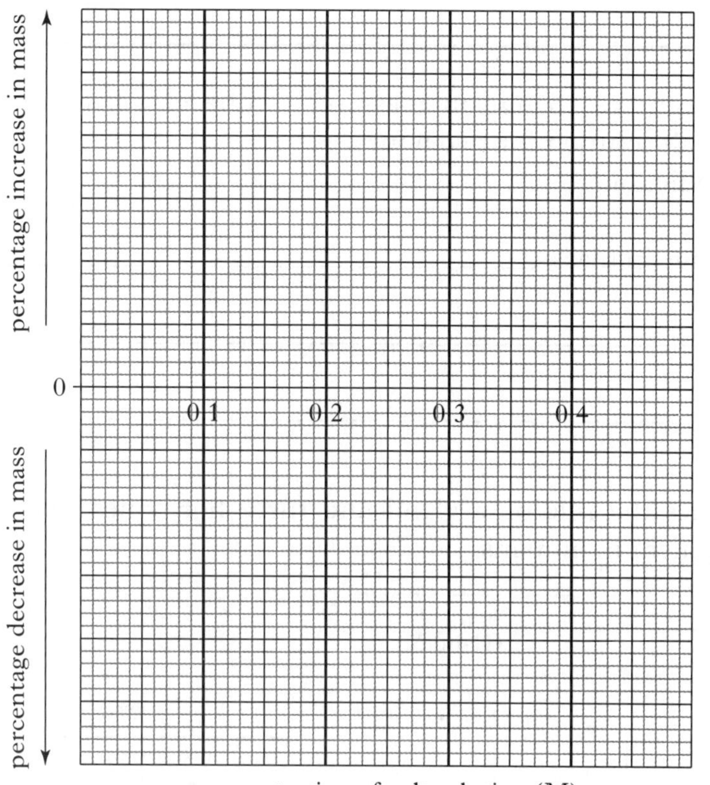

(*b*) (i) Identify a cube which is isotonic to its surrounding salt solution. Explain your choice.

Cube ____________________

Explanation ____

____ **1**

(ii) Give the terms that describe the condition of the cells of the carrot cubes after two hours in the 0.1 mol dm^{-3} and 0.4 mol dm^{-3} salt solutions.

0.1M____

0.4M____ **1**

(*c*) **From the information in the table**, explain why it is good experimental practice to use percentage change in mass when comparing results.

____

____ **1**

(*d*) (i) Identify two variables, not already described, that should have been controlled to ensure the experimental procedure was valid.

Variable 1 ____

Variable 2 ____ **1**

(ii) State one way in which the experimental procedure could be improved to increase the reliability of the results.

____ **1**

(*e*) Predict the percentage change in mass of a cube of carrot placed in a 0.05 mol dm^{-3} salt solution for two hours.
Justify your prediction.

Percentage change in mass. ______________________________%

Justification ______________________________

____ **1**

13. The diagram outlines the role of a gland and three hormones that influence growth and development in humans.

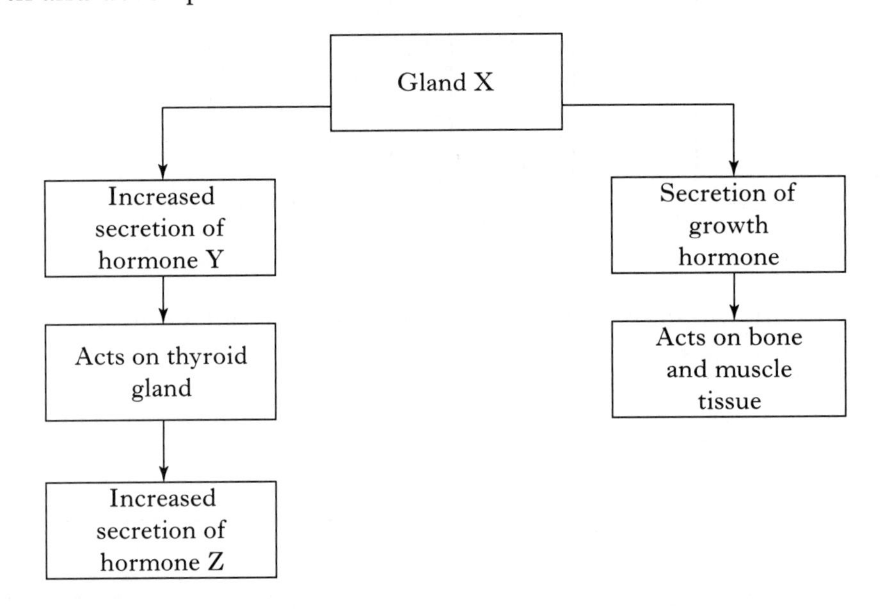

(*a*) Name gland X and hormones Y and Z.

Gland X ______________________________

Hormone Y ______________________________

Hormone Z ______________________________ **2**

(*b*) Describe the role of growth hormone in the control of growth and development.

____

____ **1**

(*c*) The change in body mass of a human male from birth to eighteen years old is shown in the graph.

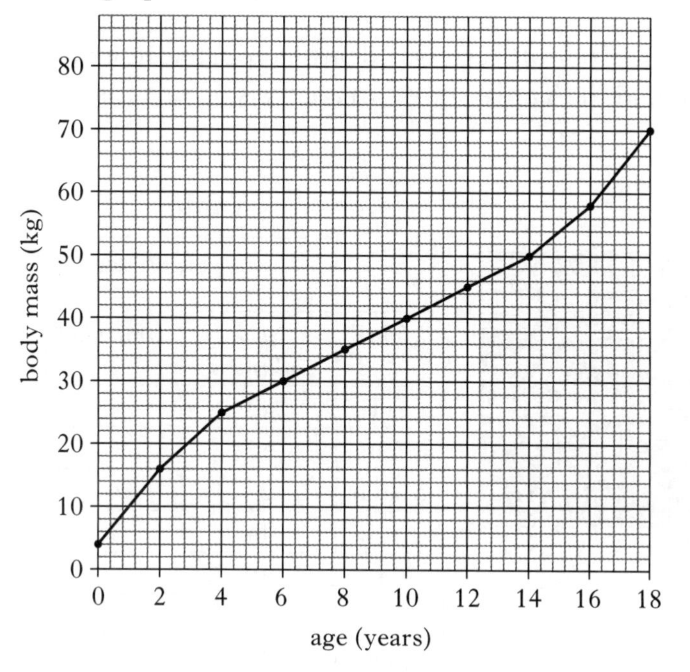

(i) Calculate the average yearly increase in body mass between birth and age 10 years.
Space for calculation

______________________ kg **1**

(ii) Tick (✓) the box to show the four year period during which the greatest increase in mass occurred.

0 – 4 years	4 – 8 years	10 – 14 years	14 – 18 years

1

14. Barn owls hunt at night but also during daylight if food is in short supply. Their food is mainly voles and shrews which live in rough grassland including roadside verges.

Eggs are laid mainly in April and May and take about a month to hatch. They are incubated by the females which remain on the nest until hatching. After the incubation period, females can leave the nest and they frequently bathe to remove soiling and parasites from their feathers. Most young have left the nest by the end of August and can hunt independently by October.

The **Chart** shows the number of barn owl deaths reported in an area of Britain over a ten-year period and the **Table** shows an analysis of various reported causes of death.

Chart

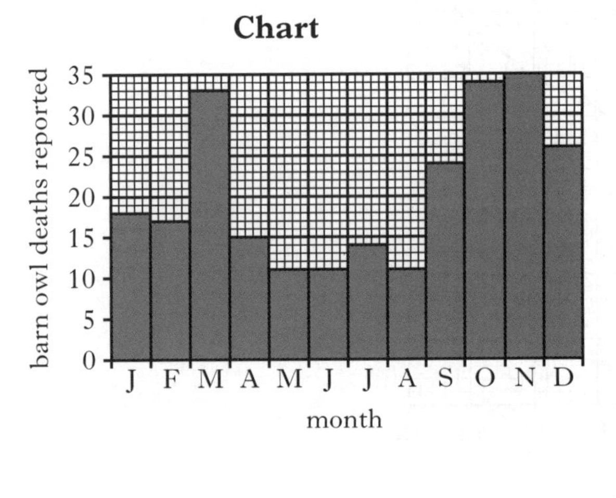

Table

	Causes of death		
Month	*Road collisions*	*Drowning*	*Starvation*
Jan	13	0	3
Feb	12	0	5
March	32	0	1
April	14	1	0
May	10	1	0
June	7	2	2
July	10	4	0
Aug	7	2	2
Sept	20	1	3
Oct	31	0	3
Nov	33	0	2
Dec	24	0	2

(*a*) Use values from the chart to describe the changes in the number of deaths between March and November.

____

____ **2**

(*b*) Explain why barn owls are often the victims of road collisions?

____

____ **1**

(*c*) Suggest two reasons to account for the changes in number of barn owl deaths between April and August.

1 ______________________________

2 ______________________________ **2**

(*d*) Account for the pattern of deaths by drowning shown in the table.

____

______________________________ **1**

(*e*) What evidence is there that some barn owl deaths in January were caused by factors other than those shown in the table?

____

______________________________ **1**

(*f*) **From the chart**, calculate the percentage decrease in deaths between March and May.
Space for calculation

______________ % **1**

(*g*) Express as a simple whole number ratio, the number of deaths due to road collisions, drowning and starvation over the period June, July and August as **shown in the Table**.
Space for calculation

________ road collisions : ________ drowning : ________ starvation **1**

SECTION C

Both questions in this section should be attempted.

Note that each question contains a choice.

All answers must be written clearly and legibly in ink on separate paper.
Labelled diagrams may be used where appropriate.

1. Answer **either** A **or** B.

A. Write notes on osmoregulation in:

(i) salt water bony fish; **5**

(ii) the desert rat. **5**

(10)

OR

B. Write notes on obtaining food under the following headings:

(i) co-operative hunting; **3**

(ii) dominance hierarchy; **3**

(iii) territorial behaviour **4**

(10)

In question 2, ONE mark is available for coherence and ONE mark is available for relevance.

2. Answer **either** A **or** B.

A. Give an account of the Jacob-Monod hypothesis of lactose metabolism in *Escherichia coli* and of the part played by genes in phenylketonuria. **(10)**

OR

B. Give an account of the effects of indole acetic acid (IAA) on plant growth and of the role of gibberellic acid (GA) in the germination of barley grains.

(10)

[End of Question Paper]

Exam 3

Biology Higher

Practice Papers
For SQA Exams

Exam 3

You have 2 hours, 30 minutes to complete this paper.

Try to answer all of the questions in the time allowed.

Write your answers in the spaces provided, including all of your working.

SECTION A

All questions in this section should be attempted.

Answers should be given on the separate answer sheet provided on page 6

1. Freshwater algal cells which actively absorb potassium ions from sea water would be expected to have large numbers of

A chloroplasts
B ribosomes
C mitochondria
D vacuoles.

2. Plant cell walls contain

A cellulose and are selectively permeable to solutes
B phospholipids and are permeable to solutes
C phospholipids and are selectively permeable to solutes
D cellulose and are permeable to solutes.

3. When a plant cell is immersed in a hypotonic solution it will

A become flaccid
B shrink
C burst
D become turgid.

4. Which of the following will occur when a photosynthesising plant in warm conditions with an adequate supply of carbon dioxide is moved from bright light to darkness?

A The concentration of RuBP in its cells will increase and that of GP will decrease.
B The concentration of RuBP in its cells will decrease but that of GP will increase.
C The concentrations of RuBP and GP in its cells will decrease.
D The concentrations of RuBP and GP in its cells will increase.

5. Which of the following processes requires flexibility and infolding of the cell membrane?

A Osmosis
B Diffusion
C Phagocytosis
D Active transport

6. When an organ is transplanted from one person to another, there is a possibility that it could be rejected.

This is because cells of the individual receiving the organ react against foreign

A antigens
B DNA
C antibodies
D RNA.

7. Which of the following pass from the light dependent stage of photosynthesis to the carbon fixation stage?

A ATP and reduced NADP
B reduced NADP and carbon dioxide
C carbon dioxide and ATP
D ATP and oxygen

8. Lysosomes are involved in the defence of the body.

They

A carry out phagocytosis and engulf bacteria
B produce antibodies to destroy viruses
C release antigens to destroy bacteria
D contain enzymes to digest bacteria.

9. An alga has cell sap with a potassium ion concentration of 0.4 mol l^{-1} and its surrounding seawater has a potassium ion concentration of 0.05 mol l^{-1}.

What is the ratio of the concentration of potassium ions in the cells compared with that in seawater?

A 0.125 : 1.0
B 8.0 : 1.0
C 1.0 : 1.25
D 1.0 : 80.0

10. Pieces of muscle tissue were measured before and after treatment with different solutions.

The results are shown in the table.

Solution added	*Initial length (mm)*	*Final length (mm)*
1% glucose	36	36
1% ATP	42	35
1% boiled and cooled ATP	40	33
1% salt solution	37	36

Which of the following statements **cannot** be concluded from the results?

A ATP is a source of energy for muscle contraction
B ATP is denatured by high temperature
C Glucose does not provide energy directly for muscle contraction
D Slight muscle contraction may be due to an osmotic effect

11. Huntington's Disease is a serious human illness. It is caused by a dominant allele and is not sex-linked.

A man heterozygous for the condition and a woman without the condition have a son.

What are the chances of the son inheriting the condition?

A 1 in 1
B 1 in 2
C 1 in 3
D 1 in 4

12. A recessive sex-linked allele for red-green colour deficiency carried on the X-chromosome of a man will be inherited by

A 50% of his male children
B 100% of his male children
C 50% of his female children
D 100% of his female children.

13. Polyploid plants

A have reduced vigour and the diploid chromosome number
B have increased vigour and the diploid chromosome number
C have reduced vigour and sets of chromosomes greater than the diploid chromosome number
D have increased vigour and sets of chromosomes greater than the diploid chromosome number.

14. Which of the following may result in the presence of an extra chromosome in the cells of a human?

A Inversion
B Non-disjunction
C Crossing over
D Independent assortment

15. Ligase is used in genetic engineering to

A open bacterial plasmids
B cut genes from a donor chromosome
C seal genes into bacterial plasmids
D remove cell walls from plant cells.

16. Sexual incompatibility between different species of plant may be overcome by

A recombinant DNA technology
B the use of mutagens
C somatic fusion of their cells
D the use of polyploid parent plants.

17. Human growth hormone can be produced by the bacterium *E. coli* using the following steps.

1 Culture large quantities of *E.coli*.
2 Insert human growth hormone gene into the plasmid DNA of the *E.coli*.
3 Cut growth hormone gene from human chromosome.
4 Extract and purify growth hormone.

The correct order for these steps is

A 3, 2, 1, 4
B 3, 1, 2, 4
C 1, 4, 3, 2
D 1, 2, 3, 4.

18. In an animal, habituation has taken place when a

A harmful stimulus no longer produces a response
B harmful stimulus always produces an escape response
C harmless stimulus always produces an escape response
D harmless stimulus no longer produces an escape response.

19. Potato production has been improved by selective breeding. The table shows the annual potato production of five countries over a five year period.

Country	*Production (millions of tonnes)*
Russia	37
Poland	25
US	23
Ukraine	18
India	17

China produces 75% more than the average of these countries.

What is the annual potato production for China in millions of tonnes?

A 18
B 36
C 42
D 48

20. The table below shows the average increase in root length recorded when samples of seedlings of the grass *Festuca ovina* from two different soils were grown in water culture medium. Half of the seedlings from each sample were grown in a culture medium to which copper salts had been added and the other half were grown in a copper-free medium.

Origin of seedling samples	*Polluted mine soil*		*Unpolluted pasture soil*	
Culture medium	Copper salts added	Copper-free	Copper salts added	Copper-free
Average increase in root length(cm)	2.7	3.0	0.7	2.8

The seedlings from both soils were inhibited by the copper.

What was the difference in the percentage inhibition between the two samples of grass seedlings?

A 10%
B 15%
C 65%
D 75%

21. The following statements relate to meristems.

1 Some provide cells for increase in diameter of stems
2 Some produce growth substances
3 They are absent from animals
4 Their cells undergo division by meiosis

Which of the statements are true?

A 1 and 2 only
B 1, 2 and 3
C 1 and 4 only
D 1, 3 and 4

22. Part of the *E.coli* chromosome is shown below

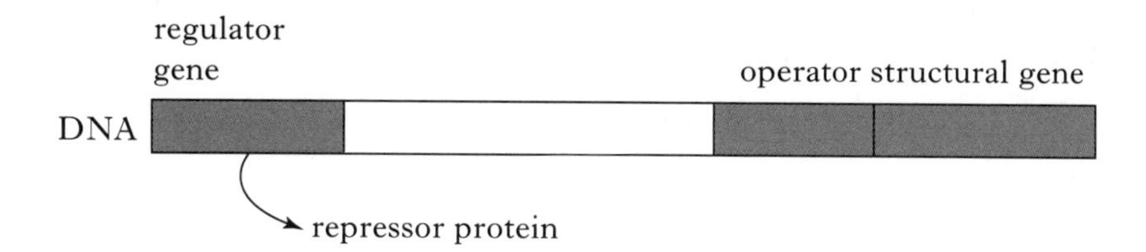

The repressor protein normally binds to the operator.

Which of the following would occur if a mutation of the regulator gene meant that the repressor could no longer bind with the operator?

A Transcription of the structural gene only when the appropriate substrate is present
B No transcription of the structural gene at any time
C Continuous transcription of the structural gene
D Intermittent transcription of the structural gene

23. Phenylketonuria (PKU) is a metabolic disorder in humans.
In PKU there is

A breakdown of excess phenylalanine in the diet
B an inability to synthesise phenylalanine
C synthesis of tyrosine from phenylalanine
D an inability to synthesise tyrosine from phenylalanine.

24. Which line in the table correctly identifies the main source of body heat and the main method of controlling body temperature in an endotherm?

	Main source of body heat	*Main method of controlling body temperature*
A	Metabolism	Physiological
B	Environment	Behavioural
C	Environment	Physiological
D	Metabolism	Behavioural

25. Which of the following are practical applications of the plant growth substance IAA (indole acetic acid)?

A selective herbicide and rooting powder
B selective herbicide and breaking bud dormancy
C fertilizer and breaking of bud dormancy
D rooting powder and fertilizer.

26. Which of the following shows the correct sequence of glands and hormones involved in the control of metabolic rate?

A TSH → thyroxine → pituitary → thyroid
B Thyroid → TSH → thyroxine → pituitary
C Pituitary → thyroxine → thyroid → TSH
D Pituitary → TSH → thyroid → thyroxine

27. Which of the following corrective mechanisms is triggered by the hypothalamus in response to an increase in body temperature?

A Relaxation of the hair erector muscles and vasodilation of the blood vessels
B Contraction of the hair erector muscles and vasoconstriction of the blood vessels
C Contraction of the hair erector muscles and vasodilation of the blood vessels
D Relaxation of the hair erector muscles and vasoconstriction of the blood vessels

28. The following graph shows the growth pattern of an annual plant.

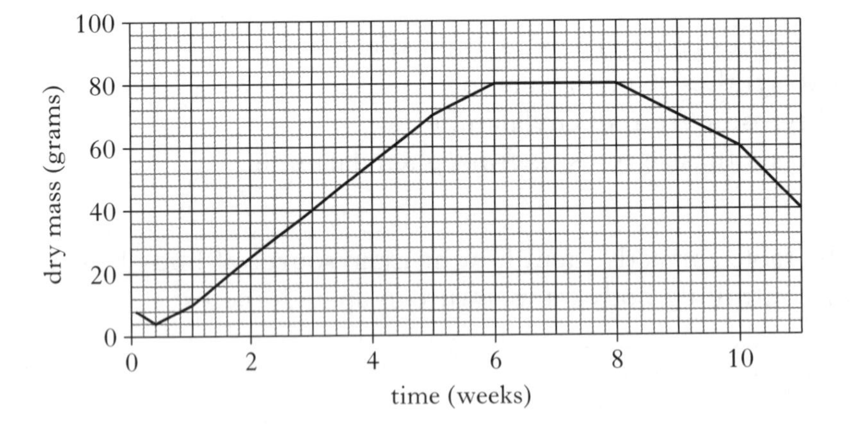

What is the percentage decrease in mass between week 8 and week 10?

A 20%
B 25%
C 60%
D 80%

29. A species of fungus is able to synthesise an amino acid essential for growth if provided with compound Q. The diagram below shows the metabolic pathway involved.

enzymes: 1, 2, 3, 4

compounds: Q → R → S → T → amino acid

A mutant strain of the fungus is found to break down compound Q when it is provided but fails to produce the amino acid and grow normally.
If provided with compound S the strain produces the amino acid and grows successfully.

This suggests that

A compound S will accumulate when compound Q is provided to the mutant strain
B the genes coding for enzymes 1, 3 and 4 are not mutated in this strain
C the strain would grow if given enough of compound R
D the gene coding for enzyme 1 is not working efficiently in the strain.

30. The graphs below show the changes in dry mass of the duckweed *L. minor* when grown alone (X) and when grown with another species, *L. gibba* (Y)

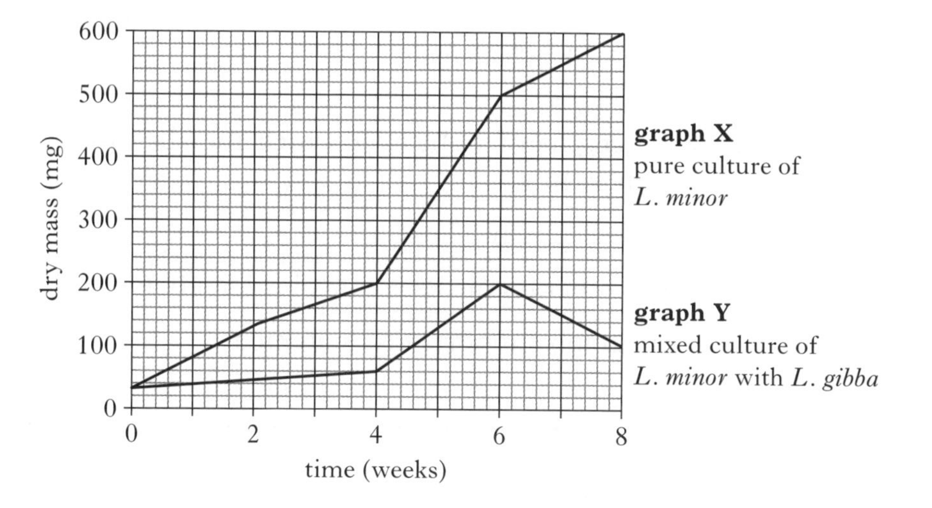

At what time was the ratio of the mass of *L.minor* grown in pure culture to that grown in mixed culture 2.5 : 1?

A 2 weeks
B 4 weeks
C 6 weeks
D 8 weeks

SECTION B

All questions in this section should be attempted.
All answers must be written clearly and legibly in ink in the spaces provided.

1. (*a*) The graph shows the absorption spectrum of the pigment chlorophyll a and the rate of photosynthesis in a green plant over the same range of wavelengths of light.

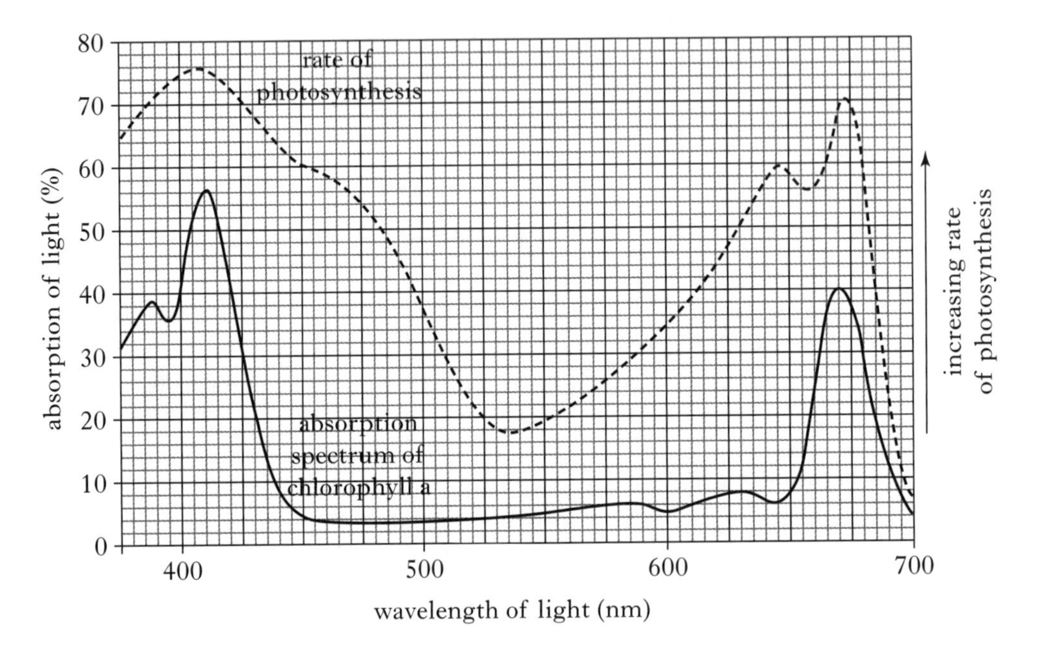

(i) Give the range of wavelengths of light over which the absorption by chlorophyll a is greater than 40%.

From ________ nm to ________ nm **1**

(ii) Apart from chlorophyll a, other pigments are involved in photosynthesis.

What evidence from the graph supports this statement?

____

____ **1**

(iii) The other pigments are accessory pigments.
State the benefit to the plant of having accessory pigments.

____

____ **1**

(*b*) (i) Apart from being absorbed, state one other possible fate of light striking a leaf.

____ **1**

(ii) Pigments that absorb light are found within the chloroplasts.
State the exact location of these pigments.

____ **1**

2. The following statements refer to the process of respiration.

1 In respiration, a high energy compound is synthesised.
2 Hydrogen released is transferred to the cytochrome system where oxygen plays its role in aerobic respiration.
3 In the absence of oxygen, the less efficient process of anaerobic respiration takes place and different end products are formed.

(*a*) Name the high energy compound. (**Statement 1**)

____ **1**

(*b*) Name the carrier that accepts and transfers hydrogen to the cytochrome system. (**Statement 2**)

____ **1**

(*c*) Describe the role of oxygen in aerobic respiration. (**Statement 2**)

____

____ **1**

(*d*) Explain why the term 'less efficient' is used in the comparison of anaerobic respiration with aerobic respiration. (**Statement 3**)

____

____ **1**

(*e*) Name the end product(s) of anaerobic respiration in an animal cell and in a plant cell. (**Statement 3**)

(i) Animal cell ________________________________

(ii) Plant cell ________________________________ **2**

(*f*) State the exact location of the cytochrome system. (**Statement 2**)

____ **1**

(*g*) In the absence of oxygen, which stage of respiration would be the only source of ATP for the cell?

____ **1**

3. (*a*) The diagram represents part of a DNA molecule undergoing replication.

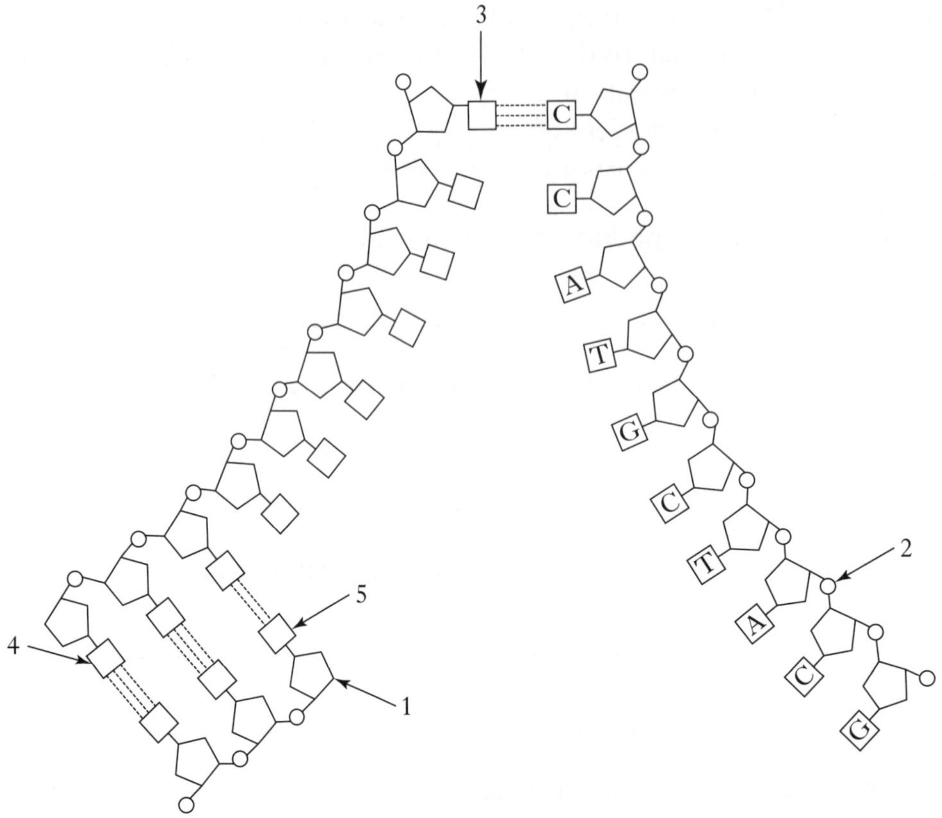

(i) Identify components 1 and 2 from the diagram.

1 ______________________________

2 ______________________________ **1**

(ii) From the diagram, identify the bases 3, 4 and 5.

Base 3 ______________________________

Base 4 ______________________________

Base 5 ______________________________ **2**

(iii) Name the type of bond which links the complementary base pairs.

______________________________ **1**

(iv) Name one other substance not shown in the diagram which is required for DNA replication to occur.

______________________________ **1**

(*b*) Explain why DNA replication must take place before a cell divides?

____

______________________________ **1**

(*c*) What name is given to the section of a DNA molecule which codes for a specific protein?

____ **1**

(*d*) The table contains information about types of protein and their function.

Complete the table. **2**

Type of protein	*Example of protein*	*Function of protein*
	Endonuclease	
Fibrous		Structural protein in skin and bone

4. (*a*) The diagram represents the structure of a virus.

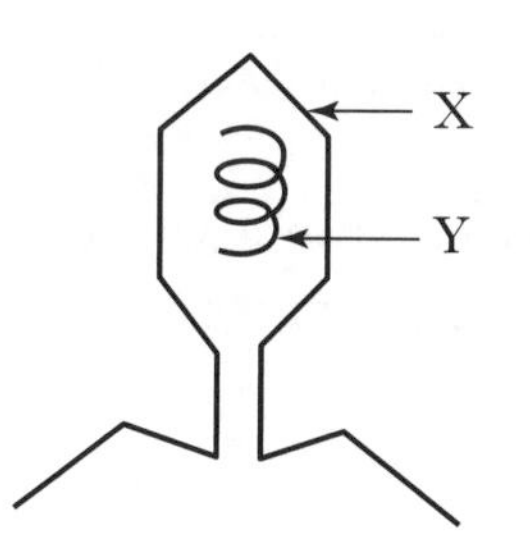

(i) Name parts X and Y.

X ____________________________________

Y ____________________________________ **1**

(ii) Name two substances, gained from the host cell, which are needed for the virus to replicate.

Substance 1 ____________________________________

Substance 2 ____________________________________ **1**

(*b*) Some plants can produce resins as a cellular defence mechanism.

(i) State one way in which resins help in plant defence.

____ **1**

(ii) Name two other chemicals produced by plants for defence.

1 ____

2 ____ **1**

5. (*a*) The graph shows the DNA content per cell nucleus present during stages of meiosis.

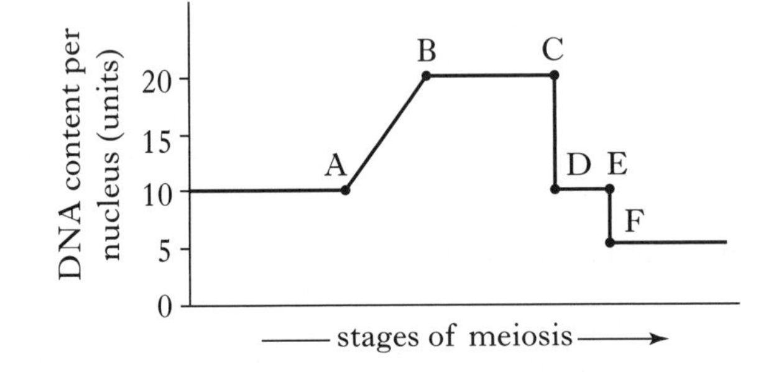

Describe what happens in meiosis which would account for the changes in DNA content between the following stages.

(i) Stages A and B.

____

____ **1**

(ii) Stages C and D.

____

____ **1**

(iii) Stages E and F.

____

____ **1**

(*b*) The diagram below shows a homologous pair of chromosomes during meiosis.

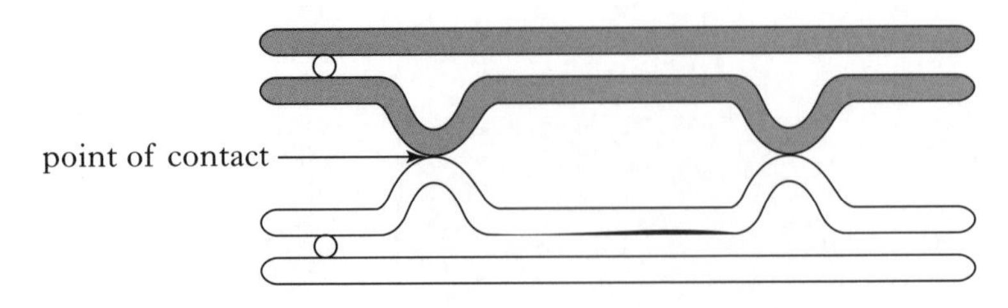

(i) What name is given to the points of contact between the chromatids of the homologous pairs of chromosomes?

____ **1**

(ii) What term describes the exchange of chromatid material between members of a homologous pair and explain how this can be a means of producing new phenotypes in a population.

Term ____________________________________ **1**

Explanation ________________________________

____ **1**

6. Exposure to radiation induces mutations in cells.

The table shows the number of mutations occurring in mouse cells exposed for different times to different radiation sources.

Length of time exposed to radiation (hours)	*Number of mutations per cell*	
	Source 1	*Source 2*
1	5	20
2	7	40
3	11	60
4	19	70
5	35	80

(*a*) Give two conclusions which can be drawn from the information provided.

1 ____

____ **1**

2 ____

____ **1**

(*b*) Calculate the percentage increase in the number of mutations per cell when the time of exposure to radiation Source 2 was increased from 1 to 2 hours.

Space for calculation

______________% **1**

(*c*) Predict the number of mutations per cell that would be expected if mouse cells were exposed to radiation Source 1 for six hours and justify your answer.

Mutations per cell ______________________________

Justification ______________________________

______________________________ **1**

7. (*a*) Salmon and eels have adaptations which maintain their water balance during their migration between sea water and the freshwater of the rivers.

The graph shows the drinking rate of a salmon which was transferred from sea water to freshwater.

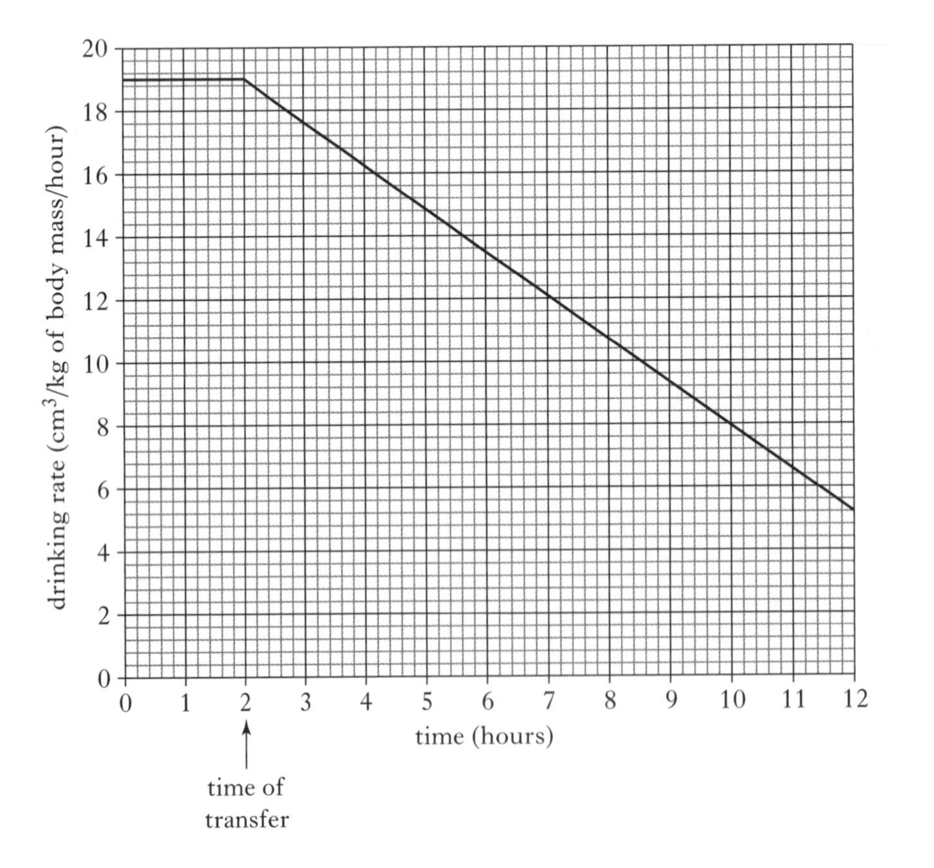

(i) Calculate the volume of water drunk by a 2.0 kg salmon in a 2 hour period before it was transferred to freshwater.

Space for calculation

______________ cm^3 **1**

(ii) In the table below tick (✓) the boxes which show the changes which occur when a salmon is transferred to freshwater.

Kidney function	*Increases*	*Decreases*	*Stays the same*
Rate of filtration			
Rate of production of urine			

1

(iii) Describe the action of the chloride secretory cells of a salmon in a sea water environment.

____

____ **1**

(*b*) Give one behavioural and one physiological adaptation shown by the desert rat to reduce water loss.

Behavioural ______________________________ **1**

Physiological ______________________________ **1**

8. (*a*) An investigation was carried out to compare the transpiration rates of Geranium and Coleus plants.

A leafy shoot was cut from a plant of each species and placed into the apparatus shown in the **Diagram**. Water loss from each shoot was measured by using the balance readings to calculate the loss in mass from each shoot and its container. Readings were taken every hour for six hours and the results shown in the **Table**.

Diagram

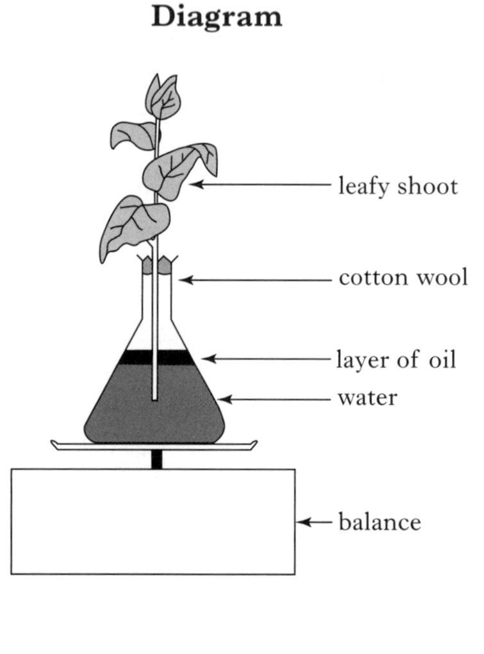

Table

Time (hours)	*Total mass of water lost* (g)	
	Geranium	*Coleus*
0	0	0
1	2	2
2	5	3
3	10	7
4	18	12
5	27	17
6	40	24

(i) Suggest a precaution which should be taken when selecting the leafy shoots for this investigation.

____

____ **1**

(ii) Explain why the layer of oil was included in the experimental design.

____

____ **1**

(iii) Suggest how the reliability of this investigation could be improved.

____

____ **1**

(iv) Identify two environmental factors which would need to be kept constant to allow a fair comparison between the two plant shoots.

Factor 1 ____

Factor 2 ____ **1**

(*b*) On the grid below, draw a line graph to show the total loss of water from the Geranium shoot over the six hours of the investigation.

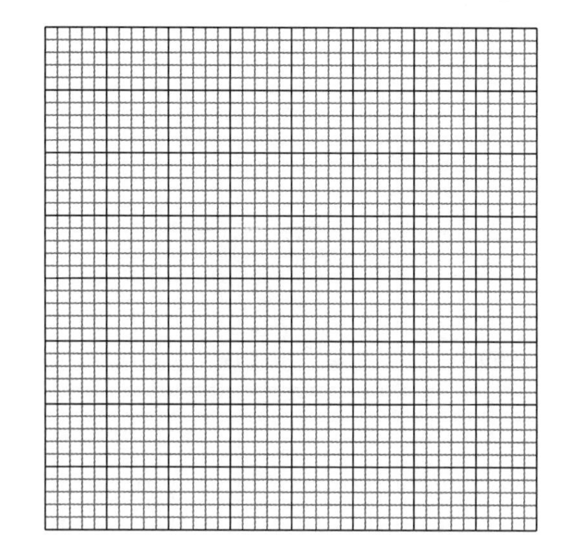

2

(*c*) Calculate the average loss of water per hour from the Coleus over the first 4 hours.

Space for calculation

________________ g/hr **1**

(*d*) State one benefit of transpiration to a plant.

____

____ **1**

9. Worker bees forage for sources of food and communicate the distance and direction of the food sources they find to other workers by dance-like movements performed after returning to the hive. They use two types of dance.

The **Table** shows the types of dances performed by workers to describe the distance and direction of the food source sites shown on the **Diagram**.

Table **Diagram**

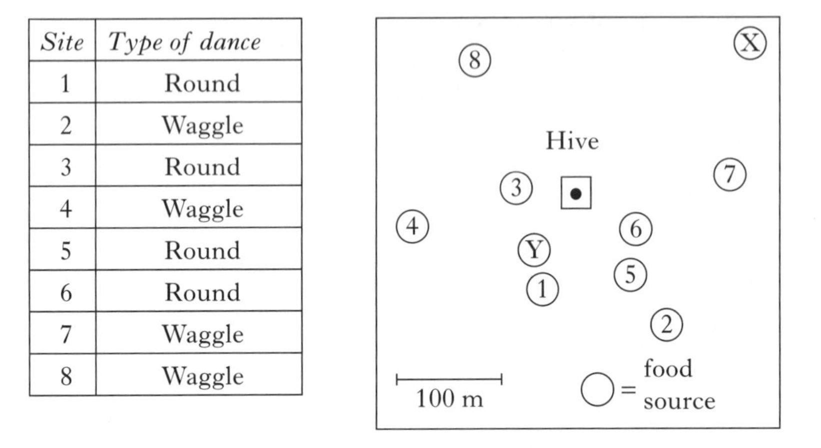

Site	*Type of dance*
1	Round
2	Waggle
3	Round
4	Waggle
5	Round
6	Round
7	Waggle
8	Waggle

Graphs 1 and 2 relate to the waggle dance. The waggle dance involves wing vibrations and is often repeated a number of times within one performance.

Graph 1 shows the relationship between the vibration of the wings and the distance to the food source.

Graph 2 shows the relationship between the number of times the dance is repeated and the distance to the food source.

Graph 1 **Graph 2**

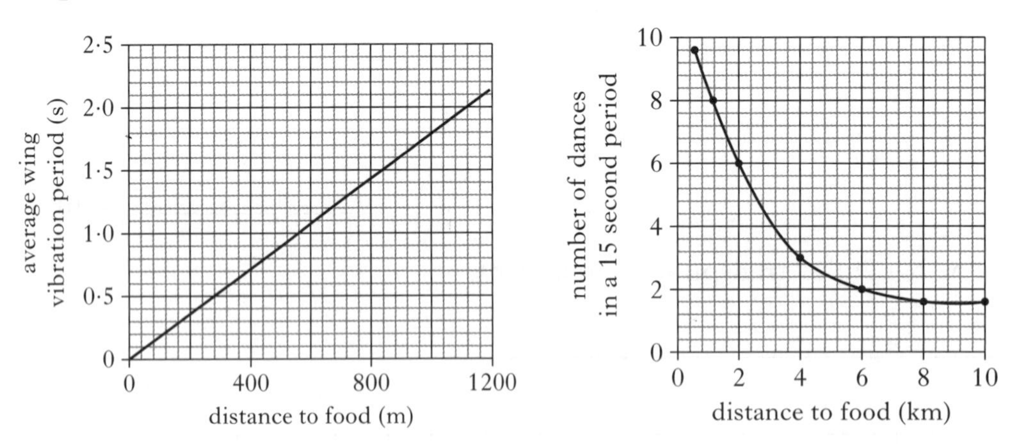

(*a*) (i) Describe the relationship between the type of dance performed and the food source.

______________________________ **1**

(ii) Predict the type of dances which would be performed to communicate information about food sources X and Y.
Justify your answers.

X ______________________________

Y ______________________________ **1**

Justification ______________________________

______________________________ **1**

(iii) Describe the relationship between the distance of the food source and

1 average wing vibration period ______________________________

______________________________ **1**

2 number of dances performed ______________________________

______________________________ **1**

(iv) State the average wing vibration and number of dances per minute that would indicate a food source located 1 km from the hive.

Average wing vibration period ______________ seconds

Number of dances ______________ per 15 seconds **1**

(v) Predict the number of dances in **one minute** if the food source was greater than 10 km from the hive.

______________ dances per minute **1**

(*b*) Explain the role of the dance-like movement behaviour in relation to the economics of foraging in bees.

____

______________________________ **1**

10. (*a*) Humming birds feed on the nectar from flowers within their territory that they defend against competitors.

The graph provides information about the territory size of a species of hummingbird in relation to the average number of flowers of its food plant present.

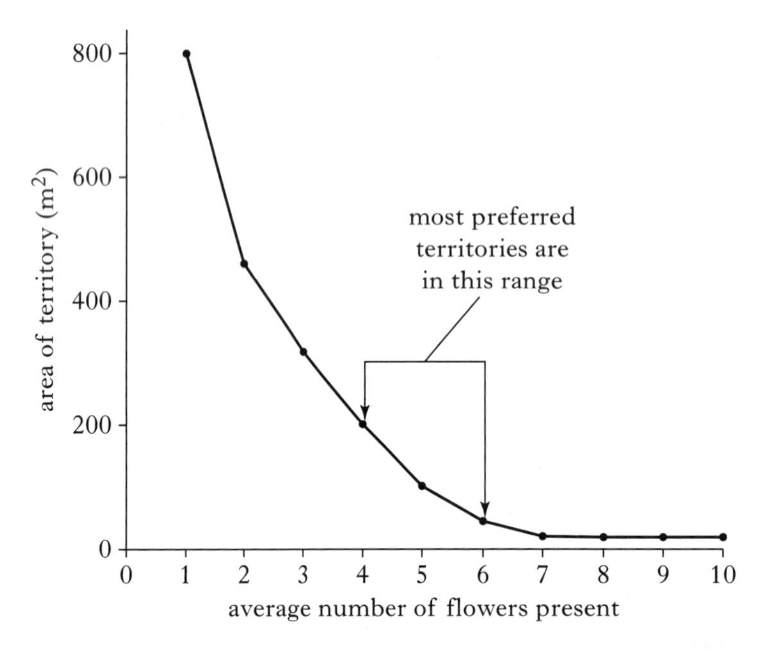

(i) Describe the relationship between territory size and flower density.

____

____

____ **1**

(ii) In terms of the economics of foraging, explain why this species prefer territories in which there are between 4 and 6 flowers per m^2.

____

____

____ **2**

(iii) Explain why an area with less than one flower per m^2 would be unsuitable in terms of a territory for hummingbirds.

____

____ **1**

(*b*) Hummingbirds chase other nectar feeding bird species and butterflies away from patches of nectar bearing flowers.

Which type of competition is reduced by chasing off other nectar feeding bird species and butterflies?

____ **1**

(*c*) Animals can economise in territorial behaviour by matching the intensity of their defence to the severity of the threat.

Defence of a territory usually occurs in a hierarchy of levels of intensity.

Intensity Level	*Example of territorial defence behaviour*
Low	Warning signal
Medium	Threat signal
High	Fighting

Explain the advantage of this hierarchy of response.

____

____ **1**

11. (*a*) The graph shows the effect of different concentrations of indole acetic acid (IAA) on the growth of plant shoots and roots.

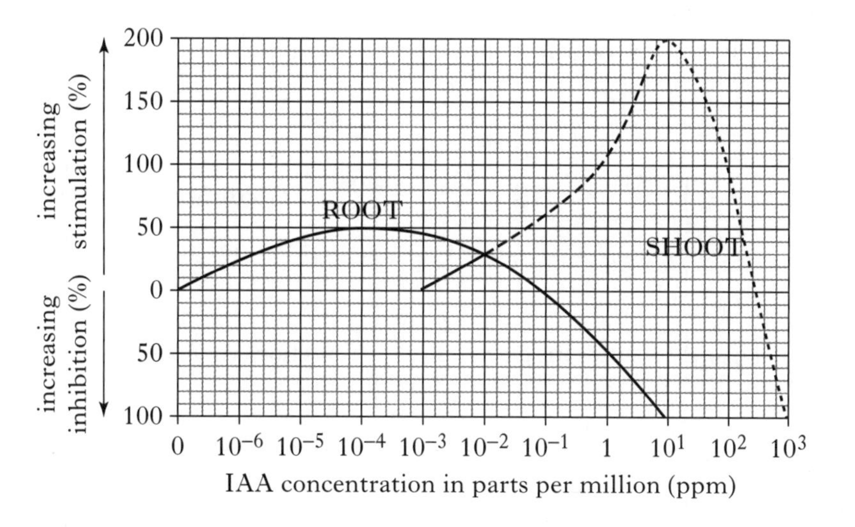

(i) Identify the range of IAA concentrations which stimulates both root and shoot growth.

From: ______________ ppm to: ______________ ppm **1**

(ii) State the IAA concentration which stimulates root and shoot growth equally.

______________ ppm **1**

(iii) Describe **using values**, the effect of increasing concentrations of indole acetic acid (IAA) on the growth of roots.

____

____ **2**

(*b*) Complete the table below by ticking (✓) the appropriate box(es) to indicate the role(s) of indole acetic acid (IAA) and gibberellic acid (GA) in the processes listed.

Process	*IAA alone*	*GA alone*	*Both IAA and GA*
α amylase production in barley grains			
Prevention of leaf abscission			
Elongation of internode cells			
Fruit formation			
Breaking bud dormancy			
Phototropism			

3

12. (*a*) A change in photoperiod can initiate flowering in plants.

Two species of plant, iris and goldenrod were exposed to different treatments of dark and light periods.

The periods of light and dark and their effects on flowering are shown in the diagram.

Photoperiod	**Iris**	**Goldenrod**
midnight 6·00 pm 6·00 am noon	flowering occurs	no flowering occurs
midnight 6·00 pm 6·00 am noon	no flowering occurs	flowering occurs

From the information given, identify which plant is a short-day plant.

Justify your answer.

Short-day plant ______________________________

Justification ______________________________

______________________________ 1

(*b*) State one way in which a change in photoperiod can affect behaviour in birds.

____ **1**

13. (*a*) The flowchart represents part of the homeostatic control of blood glucose concentration in a human.

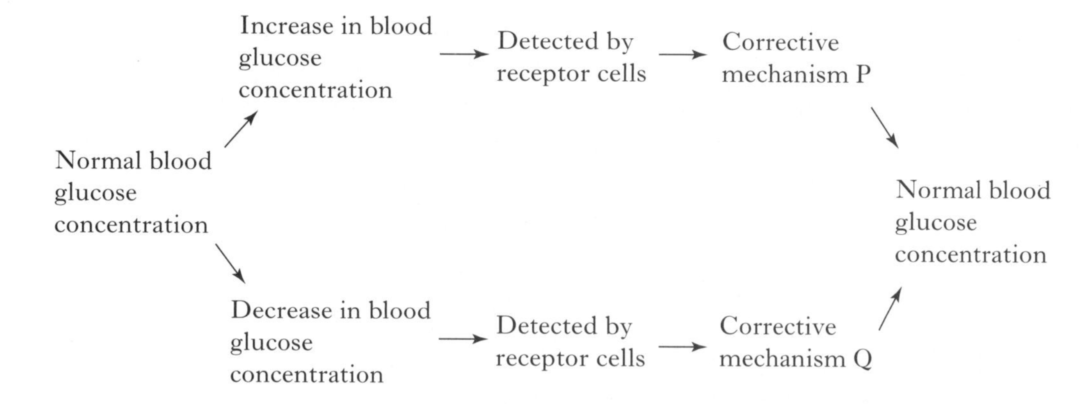

(i) State the location of the receptor cells which detect changes in the blood glucose concentration.

____ **1**

(ii) Complete each box in the table by inserting the word increases or decreases.

During corrective mechanism P	
Insulin concentration	*Glucagon concentration*

1

(iii) In corrective mechanism Q, a storage carbohydrate is broken down to release glucose into the blood.

Name the storage carbohydrate and the organ which stores it.

Name ____

Organ ____ **1**

(*b*) Control of blood glucose concentration involves negative feedback.

Explain what is meant by negative feedback control.

____

____ **2**

14. (*a*) In Britain the native red squirrel population is decreasing and facing extinction due to competition from the introduced grey squirrel. Grey squirrels out-compete the reds for food and shelter and spread the squirrel-pox virus to which they are immune but which kills red squirrels.
The graph shows the results of a study undertaken in an area of woodland to monitor the populations of grey and red squirrels.

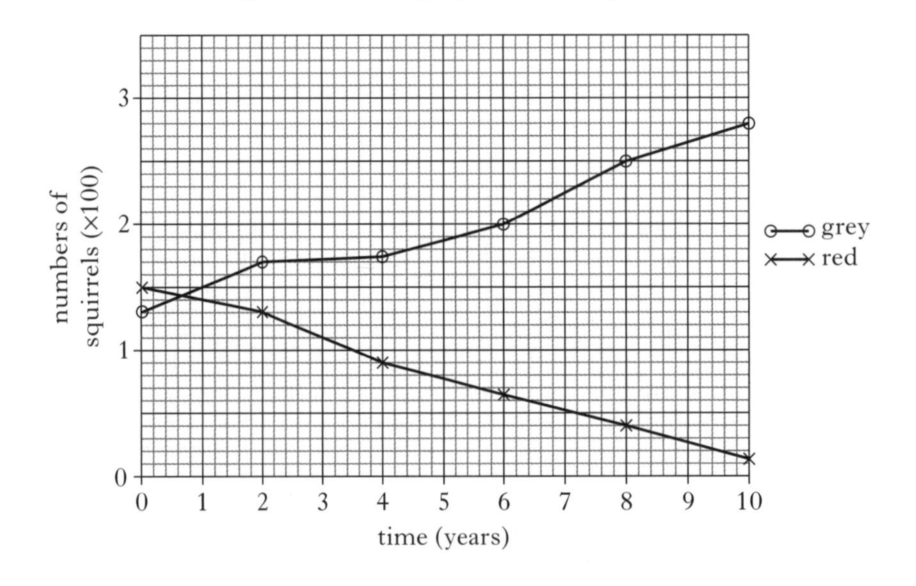

(i) Calculate how many more grey squirrels than red squirrels there were in year 8 of the study.

Space for calculation

______________________ **1**

(ii) Suggest what action might be taken to conserve red squirrel numbers.

____ **1**

(*b*) Apart from collecting data on endangered species, state two other reasons why wild populations are monitored.

(i) ____

(ii) ____ **1**

SECTION C

Both questions in this section should be attempted.

Note that each question contains a choice.

All answers must be written clearly and legibly in ink on separate paper.
Labelled diagrams may be used where appropriate.

1. Answer **either** A **or** B.

A Give an account of respiration under the following headings:

(i) glycolysis; **5**

(ii) Krebs (Citric Acid) cycle. **5**

OR **(10)**

B Give an account of protein synthesis under the following headings:

(i) structure of mRNA **3**

(ii) the roles of mRNA and tRNA in protein synthesis **7**

(10)

In question 2, ONE mark is available for coherence and ONE mark is available for relevance.

2. Answer **either** A **or** B.

A Give an account of the importance of nitrogen in plant growth and the importance of vitamin D and iron in human growth.

OR **(10)**

B Give an account of the influence of density dependent factors on animal populations and of succession in plant communities.

(10)

[End of Question Paper]

Worked Answers

EXAM 1 WORKED ANSWERS

SECTION A

Question	Answer	Helpful tips
1	A	Learning the carbon numbers is easy but essential!
2	D	Remember that if the line is sloping upwards the limiting factor is on the x axis – if the line is level the factor is **not** on the x axis. The question stem indicates that light intensity is high and has a vital clue because 10°C is quite low.
3	C	Although glycolysis is anaerobic itself, it is also the anaerobic **phase** of aerobic respiration – a fact essential to answering this question.
4	B	Acetyl groups derived from pyruvic acid are carried by CoA and join with a 4 Carbon compound to form citric acid in Krebs cycle.
5	C	The globular proteins include, some hormones, enzymes and antibodies, the fibrous ones are structural and include collagen.
6	B	Although only partly needed here, candidates are expected to know that tRNA picks up specific amino acids and carries them to the ribosomes for synthesis into protein.
7	A	When referring to nucleic acids in the context of viruses, always be clear that you are referring to **viral** nucleic acid.
8	C	The **t**hree **n**asty **c**hemicals are **t**annins, **n**icotine and **c**yanide but resin is the sticky material that isolates infected areas of certain plants.
9	B	These types of question need concentration – they are not that difficult but it is better to write down your answers to the different stages of the calculation before bringing the final answer together.
10	C	The secret is to remember that diffusion occurs from higher concentration to lower concentration and active transport is the opposite. Try each option to see if it works – this will lead to the answer.
11	A	Use rough paper to work out the options – a tricky question! Remember, sex-linked alleles are on the X chromosome.
12	D	Many candidates confuse gene and chromosome mutations – you just have to learn them but remember that deletion and inversion are terms used in **both**!
13	A	You would have to work through each set of steps eliminating the wrong ones. Hopefully the answer remains!
14	C	Many candidates confuse toleration with **avoidance**! Avoidance involves discouraging grazing using spines, thorn and stings etc. Toleration involves recovering quickly after being grazed by having underground or low growing points which re-grow damaged parts after grazing.
15	B	To answer this one you need to understand that the compensation point is the light intensity at which the carbon dioxide made by respiration exactly equals the carbon dioxide absorbed by photosynthesis.

SECTION A (continued)

Question	Answer	Helpful tips
16	D	Straightforward averages but read the question carefully and make sure your calculator works!
17	D	The secret is to fully understand the graph – what does it actually show?
18	C	The further apart genes are on a chromosome, the more frequently they are recombined because there is more space between them in which a chiasma can form. You should do a little drawing on the scrap paper to help work out the answer!
19	B	Xerophytic adaptations cut down water loss and allow survival in habitats in which liquid water is scarce.
20	A	Somatic fusion is the fusion of somatic cells of plants of different species which are sexually incompatible and cannot be crossed normally.
21	B	Just learn the graphs for annual plant, perennial plant, locust and human – they are the only ones you need to know.
22	A	Remember that thyroxine controls the metabolic rate and usually **more** thyroxine means **increased** metabolic rate.
23	C	Remember, in percentage change – find the change then divide the change by the original figure.
24	B	You must learn which responses by skin tend to release heat and which tend to conserve heat.
25	A	Endotherms and ectotherms – endo means within and ecto means outside – this is where the heat comes from in each case.
26	D	Part of the knowledge needed for this question is that a population is the number of members of the **same** species in a habitat.
27	D	Take each possibility and try to confirm it from the graph.
28	B	An absolutely standard experimental situation that you must be familiar with.
29	C	A tricky one – go back to the question and pencil in what would happen to Q and R with no enzyme 2.
30	D	Start off by looking at tissue K – is it xylem or cambium? If you know that it must be xylem the rest should be easy!

SECTION B

Question 1

Question	**Answer(s)** **/separates alternative correct answers**	Mark	Helpful tips
1 (*a*) (i)	20%	1	Percentage change calculations require two steps. In the first, the actual change has to be found – in this case a loss of 1.5g. The second step is to divide the change by the starting value and multiply the answer by 100.
1 (*a*) (ii)	5.6g	1	In making predictions from graphs, the trend of the last two points should be extended. A ruler is essential for this skill!
1 (*b*) (i)	Plasmolysed/plasmolysis	1	Just needs learning!
1 (*b*) (ii)	Water has moved from a high water concentration inside the cell to a lower water concentration outside the cell **AND** Causing the cell contents/cytoplasm and vacuole to shrink	1 1	Explanations can require several points to gain the full mark – here there are two main points. Note that for the first point the direction of the water movement is crucial to gaining the mark.
1 (*b*) (iii)	Cellulose (fibres)	1	It is worth remembering that cellulose in the cell wall is in the form of **fibres** – although this point is not needed here.

Question 2

Question	**Answer(s)** **/separates alternative correct answers**	Mark	Helpful tips
2 (*a*) (i)	X oxygen Y inorganic phosphate/Pi both needed	1	The secret here is to concentrate on what is happening to the energy which starts off trapped in glucose and is eventually used for muscle contraction.
2 (*a*) (ii)	Transfers energy from respiration to energy requiring activities in cells/muscle contraction	1	The key to this answer is the word "transfer" – many candidates will miss this vital point out of their answers.
2 (*b*) (i)	Crista/cristae	1	The cytochrome system is linked to these membranes.
2 (*b*) (ii)	Matrix (of mitochondria)	1	It is better to say "matrix of mitochondrion" – that's not needed here because mitochondrion is mentioned in the question.

SECTION B (continued)

Question 3

Question	Acceptable Answer(s) /alternative correct answer	Mark	Helpful tips
3 (*a*) (i)	Increases from 0 to 62 units as oxygen increased from 0 to 3% **AND** Then stays steady at 62 units as oxygen increased from 3 to 4%	1	You must give values from the graph to get the marks so always quote both from the X and Y axes and include the important points where change occurs.
3 (*a*) (ii)	3 times	1	Read the graph carefully and write down the figures in the space – then do the final calculation
3 (*b*) (i)	Increase in oxygen results in increased uptake of sodium	1	Remember that, generally, the factor on the X axis scale is limiting the Y axis process if increases in the factor also result in increases in the process.
3 (*b*) (ii)	A factor other than oxygen concentration is limiting the rate of uptake **OR** Sodium ion uptake is at its maximum rate	1	If a graph is level then increases in the factor on the X axis are having no effect on the process on the Y axis and the process must be limited by another factor. It is also possible that the Y axis process is at its maximum!
3 (*c*)	1.5% oxygen	1	Find the uptake of sodium at 30°C from Graph 2 – this is 50 units per minute. From Graph 1 read across from the 50 units per minute point to the graph then down to find the oxygen concentration – tricky!
3 (*d*)	1:5	1	Easy ratio – just read the figures carefully.
3 (*e*)	Active transport requires ATP from respiration Respiration is enzyme controlled Enzymes affected by temperature	All three = 2 marks Two = 1 mark	You need to notice that this graph shows the typical effects of temperature on **enzyme** activity and link this with the fact that respiration is enzyme controlled. Respiration is needed to produce the ATP needed to drive active transport.

SECTION B (continued)

Question 4

Question	Acceptable Answer(s) /alternative correct answer	Mark	Helpful tips
4 (*a*) (i)	Nucleotide	1	Deoxyribose + phosphate + base = nucleotide!
4 (*a*) (ii)	Hydrogen bond	1	It is the weakness of these bonds that allow the ease of unzipping of the strands
4 (*b*)	360	1	The base pair rules allow this calculation – adenine **must always** pair with thymine and guanine must always pair with cytosine.
4 (*c*)	DNA replication two new strands are being formed	1 1	Remember that replication is very similar to transcription – make sure you learn up the differences

Question 5

<table>
<tr><th>Question</th><th>Acceptable Answer(s) /alternative correct answer</th><th>Mark</th><th>Helpful tips</th></tr>
<tr><td>5 (a) (i) and
5 (a) (ii)</td><td>
<table>
<tr><td></td><td></td><td colspan="4">Male gametes</td></tr>
<tr><td></td><td></td><td>GW</td><td>Gw</td><td>gW</td><td>gw</td></tr>
<tr><td rowspan="4">Female gametes</td><td>GW</td><td>GGWW</td><td>GGWw</td><td>GgWW</td><td>GgWw</td></tr>
<tr><td>Gw</td><td>GGWw</td><td>GGww</td><td>GgWw</td><td>Ggww</td></tr>
<tr><td>gW</td><td>GgWW</td><td>GgWw</td><td>ggWW</td><td>ggWw</td></tr>
<tr><td>gw</td><td>GgWw</td><td>Ggww</td><td>ggWw</td><td>ggww</td></tr>
</table>
</td><td>1
1</td><td>Practice this type of grid for a cross with two heterozygotes – it's an important idea. In this question two offspring phenotypes are given so they are where you must start. It is one mark for the gametes and another for the offspring. You could still get a mark for offspring which match incorrect gametes!</td></tr>
<tr><td>5 (b)</td><td>9 grey normal: 3 grey vestigial: 3 black normal: 1 black vestigial</td><td>1</td><td>The typical ratio for dihybrid cross when two heterozygotes are crossed– but remember ratios are not always standard!</td></tr>
<tr><td>5 (c)</td><td>The ratio would be about 3 grey normal: 1 black vestigial
AND
there would be no/smaller numbers of recombinant types</td><td>1

1</td><td>Dihybrid crossed with linked genes give 3:1 ratios of parental types in the offspring – observing this ratio would indicate linked genes.</td></tr>
</table>

SECTION B (continued)

Question 6

Question	Acceptable Answer(s) /alternative correct answer	Mark	Helpful tips
6 (*a*)	Deletion would have changed all triplets after the mutation and so many amino acids would be affected	1	Remember that substitution and inversion are **point** mutations which cause the type of change in the question. Deletion (and insertion) are **frame shift** mutations and can affect many amino acids in a sequence.
6 (*b*)	...CCG .. GUG .. GAG...	1	Some candidates forget that RNA never contains thymine (T) but has uracil (U) instead. Remember, if you are asked to **name** a base, letters will **not** do!
6 (*c*)	Chemical agent(s) **OR** Irradiation	1	You could name chemicals or types of radiation but they would have to be correct. Chemical agents include colchicine and mustard gas, irradiation could be UV light or X rays.
6 (*d*) (i)	3 → 1 → 2	1	A handy sequence to remember for an extended response question!
6 (*d*) (ii)	Sexual reproduction/fertilization OR meiosis **OR** hybridization **OR** recombination/random assortment **OR** crossing over	1	Only mutation can give entirely **new** variation.

Question 7

Question	Acceptable Answer(s)/alternative correct answer	Mark	Helpful tips
7 (*a*) (i)	They have different bill shapes adapted for different methods of feeding	1	This means that they will not be in direct competition for each other's preferred food.
7 (*a*) (ii)	That they had evolved from a common ancestor	1	Adaptive radiation **always** involves a common ancestor.
7 (*b*) (i)	Geographical **AND** reproductive	1	Remember **GERM** – **G**eographical, **E**cological, **R**eproductive **M**echanisms Do not be confused by geological – it's just wrong!
7 (*b*) (ii)	Barriers prevent gene exchange between sub-populations/ mutations affecting one sub-population reaching another sub-population	1	Most candidates know that barriers stop gene exchange but they forget that it is the different mutations **failing** to cross the barriers that is critical in speciation and evolution.

SECTION B (continued)

Question 8

<table>
<tr><th>Question</th><th>Acceptable Answer(s)
/alternative correct answer</th><th>Mark</th><th>Helpful tips</th></tr>
<tr><td>8 (a) (i)</td><td>A, D and E all needed</td><td>1</td><td>You should remember that the adaptations for salt water bony fish are opposite to those for fresh water bony fish.
Learning the meaning of hypertonic and hypotonic is vital!</td></tr>
<tr><td>8 (a) (ii)</td><td>B and C</td><td>1</td><td>Many glomeruli needed for high filtration rate and this produces high volumes of urine.</td></tr>
<tr><td>8 (b)</td><td>Adaptation<table><tr><td>Lives in moist burrows</td><td>Tick (✓)</td></tr><tr><td>Forages nocturnally</td><td>✓</td></tr><tr><td>Does not sweat</td><td>✓</td></tr><tr><td>Eats dry food to gain moisture</td><td></td></tr><tr><td>Blood has high ADH levels</td><td></td></tr></table>1 mark per correct tick, lose 1 for an error</td><td>1
1</td><td>The adaptation can be behavioural or physiological – get this and you are half way there. Remember eating does not conserve water but allows a gain of water!</td></tr>
</table>

SECTION B (continued)

Question 9

Question	Acceptable Answer(s) /alternative correct answers	Mark	Helpful tips
9 (*a*) (i)	Energy gained from prey when competition is reduced in a territory is greater than the energy required to defend the territory by calling, flying and fighting	1	The question asks that information in the question stem is used and that the explanation be in terms of the economics of foraging in other words – specifically using the idea of energy gain and loss.
9 (*a*) (ii)	Density of prey items Competition	1	Remember that **density** of prey items is different from **number** of prey items.
9 (*b*) (i)	May	1	It is a good idea to use a ruler and make marks on the actual question paper to help in getting the correct answer in questions such as this – the eye can play tricks!
9 (*b*) (ii)	55 common shrews	1	Find April, use a pencil to mark out the common shrew – get the % then apply that to the total prey item for the month.
9 (*b*) (iii)	Number – Owls need more prey items when they have young **OR** more difficult hunting in winter Proportions – affected by the number of each species available **OR** ease of hunting of a species may change	1 1	Could be other answers to this one!

Question 10

Question	Acceptable Answer(s) / separates alternative correct answers	Mark	Helpful tips
10 (*a*) (i)	A and E	1	Lactose combines with the repressor AND the enzyme, Note that if the question suggests that there could be more than one answer – the chances are there will be!
10 (*a*) (ii)	D	1	Some candidates confuse the repressor with the regulator – just needs learning.
10 (*a*) (iii)	A	1	Remember that lactose is the inducer. The whole idea on this type of control is that the bacteria will only make enzymes that it can use.
10 (*b*)	Genes control the production of specific enzymes **AND** Specific enzymes control specific reactions which make up the pathway	1 1	Again a two part answer. It is always worth remembering the specific nature of enzymes and use the idea in answers.

SECTION B (continued)

Question 11

<table>
<tr><th>Question</th><th>Aceptable Answer(s)
/alternative correct answers</th><th>Mark</th><th>Helpful tips</th></tr>
<tr><td rowspan="2">11 (a) (i)</td><td rowspan="2"><table><tr><th>Experimental procedure</th><th>Reason</th></tr><tr><td>soak the grains in water for 24 hours</td><td>Soften grains/allow soluble substances to diffuse</td></tr><tr><td>wash the grains to remove bacteria or fungi</td><td>Ensure that any amylase comes from grain not microorganisms</td></tr></table></td><td>1</td><td>The best answer is to do with mobilising substances – it is more biological!</td></tr>
<tr><td>1</td><td>Tough one – easier if you have been taught this, but you could work it out!</td></tr>
<tr><td>11 (a) (ii)</td><td>Is used as a control to show that it is the GA which is causing the effect</td><td>1</td><td>Control measures are designed to help to show what is actually causing the results.</td></tr>
<tr><td>11 (a) (iii)</td><td>Mass/concentration of GA
OR variety of barley grain</td><td>1</td><td>Tricky – most already mentioned.</td></tr>
<tr><td>11 (b) (i)</td><td>Scales and labels correct</td><td>1</td><td>Scales must always be enclosed and include zero if appropriate. Labels must be taken directly from the headings in the data table in the question.</td></tr>
<tr><td>11 (b) (ii)</td><td>Points plotted accurately, connected and lines labelled</td><td>1</td><td>All points should be plotted carefully to indicate the exact point and label lines connected using a ruler.</td></tr>
<tr><td>11 (c)</td><td>2.1 units</td><td>1</td><td>In predicting from graphs, simply extend the graph line as indicated by the trend shown over the last two points. You can also get this from the table.</td></tr>
<tr><td>11 (d) (i)</td><td>The embryo produces GA</td><td>1</td><td>Those without embyos did not produce amylase unless they were given GA.</td></tr>
<tr><td>11 (d) (ii)</td><td>GA needed for amylase production/synthesis</td><td>1</td><td>Only dishes with either GA or an embryo to produce it showed any amylase production.</td></tr>
<tr><td>11 (e)</td><td>Aleurone (layer)</td><td>1</td><td>Just needs learned.</td></tr>
</table>

SECTION B (continued)

Question 12

Question	Acceptable Answer(s) /alternative correct answers			Mark	Helpful tips
12 (*a*)	*Macroelement*	*Importance*	*Symptom(s) of deficiency*		
	phosophorus	found in DNA and ATP	reduced growth/ red leaf bases	1	Red leaf bases in cereals is characteristic of both phosphorus and nitrogen deficiency.
	potassium	needed for membrane transport	reduced growth	1	Potassium is involved in the movement of molecules across membranes.
	magnesium	found in chlorophyll	chlorosis/ description	1	You could also say that leaves would be pale or yellow.
12 (*b*) (i)	Light from one side of a growing shoot causes the shoot to grow in the direction of the light source.			1	Do not use the term "bends" towards light.
12 (*b*) (ii)	Absence of light results in a growing shoot elongating/growing long internodes **OR** having small yellow leaves.			1	This is etiolation.

SECTION B (continued)

Question 13

Question	Acceptable Answer(s) /alternative correct answers	Mark	Helpful tips
13 (*a*) (i)	Pancreas	1	The hormones are secreted from the pancreas but have their effects in the liver.
13 (*a*) (ii)	X insulin Y glucagon	1 1	Just needs learning. Be careful with spelling of glucagon – and confusion with glycogen will lose the mark!
13 (*b*)	Change in blood glucose concentration detected by receptors and triggers a corrective response Corrective response returns glucose concentration to normal level which switches off corrective response	1 1	It is worthwhile to memorise a general definition of negative feedback – it is a difficult idea.
13 (*c*) (i)	Individual Q 1 blood glucose concentrations above normal at start **OR** never returned to normal concentrations 2glucose concentration increased much more after ingestion **OR** took longer to return to pre-ingestion concentrations	1 1	You really need to use the piece of information from the question which indicates normal blood glucose concentrations.
13 (*c*) (ii)	4 mg per 100 cm^3 per minute	1	60 minutes later the glucose in 100 cm^3 has gone up to 440 mg from 200 mg so this is a 240 mg increase. Divide 240 by 60 to get the average increase per minute.

SECTION B (continued)

Question 14

Question	Acceptable Answer(s) /alternative correct answers	Mark	Helpful tips
14 (*a*) (i)	The total number of pups born between the two sites was greater than the maximum born on the Farne Islands before the exclusion	1	Relies on you understanding what the graph is showing about the seals and making a mental image of what is happening.
14 (*a*) (ii)	1980 300 more pups on the Farne Islands 1990 50 more pups of the Isle of May	1	Use a pencil to mark the relevant places on the graph.
14 (*a*) (iii)	1970–1980	1	Marking on graph will show the largest change – a decrease on the Farne Islands in this decade.
14 (*b*) (i)	Disease or food supply or predation or competition for a named resource	1	Density dependent factors have greater effects on populations as population density increases
14 (*b*) (ii)	Temperature/examples of extreme temperature or rainfall/examples of extremes of rainfall	1	Density independent factors operate equally on small populations as on large populations.

SECTION C

Question 1A

Your answer should include / separates alternative correct answers	**Mark**	**Commentary and Helpful tips**
Named pigments – chlorophyll a, b, xanthophyll, carotene,	1	**A maximum of 4 marks for this section.** You must be able to name the pigments and which regions of the spectrum they absorb. Chlorophyll a is the main pigment and the others broaden the absorption spectrum by absorbing light that chlorophyll a cannot and passing the energy they receive onto chlorophyll a.
Chlorophyll b, xanthophyll and carotene are accessory pigments	1	
Chlorophyll absorbs mainly in blue and red regions of spectrum	1	
Accessory pigments absorb in other regions of the spectrum/broaden absorption spectrum	1	
Accessory pigments pass energy on to chlorophyll a	1	
Light dependent stage occurs in the grana of a chloroplast	1	**A maximum of 6 marks for this section.** You will find that learning a simple diagram or flow chart is good for questions like this. Watch though – you must label any diagrams carefully and ensure that arrows are given in flow charts and that they are pointing in the right direction. Don't confuse NADP with NAD – the NAD**P** is the carrier in **P**hotosynthesis! Lots of candidates fail to realize that the role of the hydrogen from the light dependent stage is to be used to reduce carbon dioxide to carbohydrate.
Light energy is used to regenerate ATP	1	
ATP is produced from ADP + Pi	1	
Light energy splits water molecules	1	
The splitting of water is photolysis	1	
Water is split into hydrogen and oxygen	1	
Hydrogen is accepted by NADP	1	
NADPH is passed to the carbon fixation stage	1	
Oxygen is released from plants as a by-product	1	

SECTION C (continued)

Question 1B

Your answer should include / separates alternative correct answers	**Mark**	**Commentary and Helpful tips**
Phagocytosis involves engulfing foreign material/cells by phagocytes	1	**A maximum of 5 marks for this section.**
Pocket in membrane forms around the foreign material **OR** membrane folds around the foreign material	1	Diagrams are useful but must be carefully labelled.
Particles are finally enclosed into a vacuole	1	This is really about the activities of phagocytes – special white blood cells.
Lysosomes contain digestive enzymes	1	
Lysosomes fuse with the vacuole	1	The role of lysosomes as sources of the digestive enzymes is vital.
Enzymes added to vacuole digest particles inside	1	
Digestion products are diffused into the cytoplasm	1	
Antibodies are proteins	1	**A maximum of 5 marks for this section.**
Antibodies are produced by lymphocytes	1	
Antibodies are produced in response to antigens	1	This is really about lymphocytes – another type of white blood cell.
Antibodies are specific to the antigens which stimulate their production	1	Getting the antigen and antibody confused is a common mistake here. Antigens stimulate the production of antibodies.
Antigens are foreign proteins	1	The specificity of antibodies to antigens which stimulate them is important.
Antibodies combine with antigens	1	Be careful with the use of the term foreign – it should be used to refer to antigens.
Antibodies can render antigens harmless	1	

SECTION C (continued)

Question 2A

Your answer should include / separates alternative correct answers	**Mark**	**Commentary and Helpful tips**
Meiosis occurs in a diploid gamete mother cell	1	**A maximum of 4 marks for this section.** The language is vital; diploid, mother cell, homologous, chromosome. You must stick to talking about the first meiotic division and diagram, if labelled, will be useful.
Spindle forms	1	
Homologous chromosomes pair and line up at equator of spindle	1	
Crossing over may occur	1	
Homologous chromosomes pull apart / separate	1	
Chromosomes move towards the poles of spindle	1	
Two new spindles form	1	**A maximum of 4 marks for this section.** Again vital language now includes the terms chromatid, haploid and gamete. Talk about the second division only and label any diagrams you use.
Chromosomes line up at equators of two new spindles	1	
Chromosomes split to form two chromatids	1	
Chromatids move towards poles of new spindles	1	
Cytoplasm splits into four **OR** four new cells/gametes are formed	1	
These gametes are haploid	1	
Coherence The writing should be under **sub headings** or divided into **paragraphs**. A sub heading / paragraph for first stage and a sub heading/paragraph for second stage of meiosis. The account must be presented in a **logical** and **progressive** way. Related information must be grouped together. Information on first stage (2/3 points) and information on second stage together (2/3 points). Five points needed.	1	**Two marks available for being careful** – just split the writing clearly into sections and don't mention anything that is not about meiosis!
Relevance Must not give details not directly related to meiosis – e.g. process of mitosis. Must give at least 2/3 relevant points on first stage and 2/3 points on second stage. Five points needed. **Both must apply correctly to gain the Relevance Mark.**	1	

SECTION C (continued)

Question 2B

Your answer should include / separates alternative correct answers	**Mark**	**Commentary and Helpful tips**
Water moves from soil to root cells by osmosis	1	**A maximum of 6 marks for this section.** This is all about forces and processes moving water through the plant; osmosis, cohesion, adhesion, evaporation are the keys. When water goes from a leaf, it first **e**vaporates into the air spaces within the leaf the**n** **d**iffuses out to the atmosphere – remember **end**! Ensure that you write that water **vapour** diffuses out through stomata not simply water.
Root hair cells increase the surface area of the root in contact with the soil	1	
Water then passes across the cortex along a water concentration gradient	1	
Water passes into xylem vessels	1	
Within the xylem adhesion keeps the water column together	1	
Within the xylem cohesion keeps the water column together	1	
Water enters leaf cells from the xylem	1	
Water evaporates into air spaces between the leaf cells	1	
Water vapour diffuses from the air spaces to the atmosphere leaf through stomata	1	
Transpiration provides water for photosynthesis/photolysis **OR** transpiration provides water for turgidity	1	**A maximum of 2 marks for this section.** Any two will do.
Transpiration provides minerals to the plant	1	
Transpiration cools the plant	1	
Coherence The writing should be under **sub headings** or divided into **paragraphs**. A sub heading / paragraph for transpiration and a sub heading/paragraph for its importance. The account must be presented in a **logical** and **progressive** way. Related information must be grouped together. Information on transpiration (4 points) and information on its importance (1 point). Five points needed.	1	**Two marks available for being careful** – just split the writing clearly into sections and don't mention anything that is not about transpiration!
Relevance Must not give details not directly related to transpiration. Must give at least 4 relevant points on transpiration and 1 point on its importance. Five points needed. **Both must apply correctly to gain the Relevance Mark.**	1	

EXAM 2 WORKED ANSWERS

SECTION A

Question	Answer	Helpful tips
1	B	Good question to test your knowledge on cell structures and their functions. Learn these and be able to relate certain structures with the roles of specific cells eg. the role of lysosomes in phagocytes or Golgi in the pancreas cells.
2	C	First you have to notice that there was a decrease in the mass. Next you have to be able to identify the difference in water concentrations which would result in the movement of water from solution Y out from the bag and into the surrounding solution X. The 5% salt solution in Y has 95% water and the 10% salt solution has only 90% water and so water moves out.
3	A	Remember **ART**, **A**bsorption, **R**eflection and **T**ransmission as the fates of light striking a leaf.
4	D	Should be a straightforward substitution of the values and then dividing the distance travelled by the pigment by 9, the distance travelled by the solvent front. Note carefully the position of the solvent front!
5	C	Work through the answers and eliminate each in turn based on your knowledge.
6	C	Krebs cycle takes place in the matrix of the mitochondria.
7	B	Collagen is the only fibrous protein specified in the Arrangements document. The others listed here are all globular proteins. Watch out though – cellulose is found as **fibres** but is not even a protein!
8	A	Remember that genes code for proteins eg. enzymes that control all the cells activities.
9	A	This question requires you to know that DNA is double stranded linked by complementary base pairs and composed of nucleotides with the sugar deoxyribose. There are 30 guanine nucleotides and so another 30 cytosine nucleotides Similarly with 15 adenine nucleotides there has to be 15 thymine nucleotides. This gives 90 in total each with a deoxyribose sugar molecule.
10	D	The cellular defence chemicals detailed in the Arrangements are tannins, nicotine, cyanide and resin. Check out the role of lymphocytes in the production of antibodies in animals only!
11	A	This gives the closest ratio to the expected 9 : 3 : 3 : 1
12	B	Chiasmata are the points at which members of homologous pairs of chromosomes touch and crossing over takes place ie. exchange of genetic material .
13	C	Working out this cross should give you a 1:1:1:1 ratio and from this you should notice that only one quarter of the 64 offspring should have the tall and smooth characteristics. Do working on scrap paper.
14	A	Straightforward definition but a difficult concept.
15	D	Recombination frequencies refer to the percentage crossing over that takes place between genes. The further apart the greater the frequency.

SECTION A (continued)

Question	Answer	Helpful tips
16	D	Mitosis produces identical daughter cells with no genetic variation at all!
17	C	Difficult two step question. First you need to work out the **complementary** strand of DNA **and then** look for the inversion mutation.
18	B	Speciation is the formation of two or more species from the original one.
19	D	The dark or melanic forms are now better camouflaged and so are being eaten less frequently than the lighter variety.
20	A	This question requires you to know how **changing** each of these factors affects transpiration.
21	D	This is a description of an etiolated plant.
22	A	You should try to learn a good definition for both the long day flowering plants and the short day flowering plants in terms of **both** light **and** dark hours.
23	B	If you cannot remember all of the macro-elements, why they are needed and their deficiency symptoms then you may need to try to eliminate some answers or focus in on some of the information you are more sure of eg. ribulose 1-5 bisphosphate must contain phosphorus and magnesium is required for chlorophyll.
24	C	First calculate the increase then divide it by the original value before multiplying by 100 to turn it into a percentage. In this case an increase of 1 divided by $0.5 \times 100 = 200\%$
25	C	Insulin decreases blood glucose concentration while glucagon increases it.
26	D	Less ADH makes tubule less permeable and so there is less water reabsorption which results in increased volume of dilute urine
27	C	Divide 2.1 by 0.03
28	B	Requires you to select the appropriate information and draw the correct conclusion. The concentration asked about caused the shoot to grow more than normal/than the control and the root to be inhibited.
29	B	Work out the number of seeds per plant first. Decrease in seed production as the density increased suggests increased competition has affected yield.
30	B	As succession progresses from the pioneer community through to the climax community the biomass, species diversity and food web complexity all increase.

SECTION B

Question 1

Question	Answer(s) **/ separates alternative correct answers**	Mark	Helpful tips
1 (*a*) (i)	P – Stroma Q – Carbon dioxide	Both for 1	Careful with your spelling here! St**r**oma <u>not</u> stoma. Think! **Carbon** dioxide for the **carbon** fixation stage.
1 (*a*) (ii)	Water	1	Water is hydrogen oxide (H_2O)
1 (*b*)	Transfers (chemical) energy (from the light dependent stage) to the carbon fixation stage / Calvin cycle. **OR** Provides energy for the carbon fixation stage / Calvin cycle. **OR** Provides energy for the synthesis of glucose. **OR** Provides energy for converting GP to glucose	1	Although fairly straightforward knowledge, many candidates do not give the necessary detail in the answer and stop at "ATP provides energy". This is not enough on its own.
1 (*c*)	Proteins / fats / nucleic acids	1	Although really a C type question these examples of other major biological molecules are often overlooked in this section.
1 (*d*)	1. RuBP decreases because it is converted to EP 2. It cannot be regenerated from EP 3. ATP/H_2 not available	All 3=2 1/2 correct = 1	This question is an "explain" question – much more demanding than a simple describe question. You are required to select the appropriate information from the graph, describe the trend, and then provide information to **explain** the change.

SECTION B (continued)

Question 2

Question	Answer(s) / separates alternative correct answers	Mark	Helpful tips
2 (*a*)	Glucose 6 Pyruvic acid 3 Acetyl group 2 Citric acid 6	2 (all correct 2, three or two correct,1)	Just needs learning but think about how the lost carbon dioxide molecules affect the remaining carbon number.
2 (*b*)	Fats / protein	1	Although really a C type question, the term alternative respiratory substrate and these examples are often skimmed over and forgotten
2 (*c*)	ATP	1	Remember, ATP is needed to "kick start" glycolysis!
2 (*d*)	NAD	1	If you are asked for the name of the carrier molecule, avoid giving its reduced form NADH – this is just wrong.
2 (*e*)	Acts as the final hydrogen acceptor **OR** Combines with hydrogen to form water **OR** Accepts / removes hydrogen from the cytochrome system	1	Best advice for this one is to ensure all points are covered and give the following answer: Oxygen acts as the final hydrogen acceptor and joins with the hydrogen to form water

SECTION B (continued)

Question 3

Question	Answer(s) / separates alternative correct answers	Mark	Helpful tips
3 (*a*)	X – mRNA / messenger RNA Y – tRNA / transfer RNA	1 (Both required)	This is an example of a C type question made more difficult by requiring two answers.
3 (*b*)	Ribosome	1	Remember, the ribosome is the site of protein synthesis.
3 (*c*)	1 - uracil 8 - uracil 15 - guanine	2 (All correct 2, one or two correct, 1)	Candidates often have difficulty with this type of question. Best advice for this one is to write out the bases of the DNA then the mRNA and then the tRNA. Use scrap paper if necessary and take great care when working out the complementary bases. Remember that neither mRNA nor tRNA have thymine in their molecules - thymine is replaced by uracil.
3 (*d*)	Packages / modifies / processes proteins ready for secretion. **OR** Produces vesicles containing proteins for secretion	1	The word secretion alone is **not** enough.
3 (*e*)	4	1	This question requires you to know the complementary base pairs as they apply to DNA, mRNA and tRNA. It is a good idea to draw three lines on scrap paper to represent the DNA, the mRNA and the tRNA strands. You can then write in the bases and work out the complementary base pairs for each strand in turn. Remember that uracil replaces thymine in both mRNA and tRNA.
3 (*f*)	22	1	Since the DNA contains 28% Adenine there must be 28% thymine adding up to 56%. That means that the remaining 44% of the DNA has to be composed of the other two base pairs which means there is 22% of cytosine and 22% of guanine.

SECTION B (continued)

Question 4

Question	Answer(s) / separates alternative correct answers	Mark	Helpful tips
4 (*a*)	Stage 4 - Synthesis of viral protein coat.	1	This type of sequencing question involving a flowchart of viral replication is asked fairly regularly – the idea can be used in answering extended response questions on viruses.
	Stage 6 - Lysis / host cell bursts releasing new copies of virus / new copies of virus exit by leakage	1	Take care if you are asked what happens following the injection of the viral DNA into the host cell. Many candidates lose this mark by saying that the virus replicates instead of saying the viral DNA replicates.
4 (*b*)	Nucleotides / ATP / enzymes	1	Remember that the virus has nothing but nucleic acid and a coat. The host cell provides nucleotides for the new viral nucleic acid, ATP for energy and enzymes to control the process.
4 (*c*)	2000 viruses	1	First cell releases 100 viruses. 20% of 100 is 20, which then infect 20 new cells which go on to release 100 new viruses each. This gives a total of 2000 (20×100).
4 (*d*) (i)	A resin	1	Remember that resin acts as a sticky barrier substance to prevent the entry and spread of infection to other parts of the plant.
4 (*d*) (ii)	B Antibody	1	Lymphocytes make antibodies in response to foreign antigens which may be on transplanted tissue.
4 (*e*)	Suppressor drugs / immunosuppressor drugs / drugs that suppress the immune system	1	Just needs learning!
4 (*f*)	The secondary response is quicker / faster Concentration of antibodies produced is greater Response lasts for longer.	2 (All three correct 2: One or two correct 1)	Selecting three pieces of information increases the demand of this question. Remember to use the terms on the axes labels such as concentration rather than using the term amount. Looking at the steepness of the slopes against the x-axis gives the additional answers relating to time.

SECTION B (continued)

Question 5

Question	Answer(s) / separates alternative correct answers	Mark	Helpful tips
5 (*a*)	Cross 1 - X^rY; X^RX^r Cross 2 - X^RY; X^rX^r	1 1	Sex linkage problems are often answered poorly. Remember to write the sex chromosomes XY and XX down first, then add the sex-linked alleles. Only the X chromosome carries sex-linked alleles and so the Y should not be given any additional letters. The letters used to represent the sex linked alleles must be written as superscript and not side by side next to the sex chromosomes.
5 (*b*)	Cross 2 The table shows that all females are red-eyed and all males are white-eyed	1	As seen from the table, red-eyed flies are always female and white-eyed flies are male.
5 (*c*)	Genes found on the part of the X chromosome which is not homologous to the Y chromosome	1	The demand of the answer scheme makes this an A type question. It is not enough to say that "It is a gene found on the X chromosome". You must make the comparison between the X and the Y chromosome. A little drawing might help.
5 (*d*)	Females must receive two copies of the allele (to show the condition) whereas males require only one.	1	Once again this is an example where an exact and detailed definition is required to obtain the mark. Again you need the comparison to be made between the female and male.

Question 6

Question	Answer(s) / separates alternative correct answers	Mark	Helpful tips
6 (*a*)	The chromosomes cannot form homologous pairs. This means that meiosis cannot take place and so gametes cannot be produced	1	Learn this response and be able to produce it in similar questions.
6 (*b*)	Complete non-disjunction	1	Only complete non-disjunction can produce polyploidy.
6 (*c*)	Increased vigour / resistance to disease / resistance to drought / increased fruit size or crop yield.	1	Whenever asked about an improvement or an advantage in a question like this you need to include the term "increased" in the answer

SECTION B (continued)

Question 7

Question	Answer(s) / separates alternative correct answers	Mark	Helpful tips
7 (*a*)	Geographical	1	Take care. For some reason, candidates often use the term geological rather than geographical. Remember! The role of an isolating barrier is to act as a barrier to gene exchange or gene flow
7 (*b*)	Common ancestor **OR** Evolved from same species. Different beak size or shape to feed on different foods. **OR** Adapted to fill different ecological niches.	1 1	Adaptive radiation questions require the candidate to refer to a common ancestor and then to describe how each species has evolved and adapted. Note that in this particular example, the mark would not have been awarded for simply saying that they had different beaks. You needed to mention in what way they were different and why.
7 (*c*)	Alternative food supply increases chance of survival. **OR** If one food is in short supply then they can feed on another	1	More difficult question requiring you to make links or connections to suggest a suitable explanation.

Question 8

Question	Answer(s) / separates alternative correct answers	Mark	Helpful tips
8 (*a*)	Gene probe / recognition of chromosome banding pattern	1	These are the options suggested by the Arrangements – safer to use them.
8 (*b*)	Endonuclease Endonuclease Ligase	All 3 = 2 ½ = 1	The same general type of enzyme is used in both steps 3 and 4.

SECTION B (continued)

Question 9

Question	Answer(s) / separates alternative correct answers	Mark	Helpful tips
9 (*a*)	X - Decrease in mass as seed uses up food store / starch to grow / germinate / in respiration Y - Increase in mass due to photosynthesis. Z - Decrease in mass due to seed / fruit dispersal.	All 3 = 2 1 or 2 = 1	This requires full selection and explanation, using knowledge, of the trends shown in the graph.
9 (*b*) (i)	Name - Cambium. Function - Region of mitosis / cell division. **OR** Produces new cells.	1 1	Remember! Mitosis produces identical daughter cells, so when the cambium divides it must produce more cambium cells first which may then undergo differentiation to form xylem and phloem.
9 (*b*) (ii)	C Larger diameter	1	Really all you need to know is that spring wood has vessels with a wider diameter than those of the summer wood!

SECTION B (continued)

Question 10

Question	Answer(s) / separates alternative correct answers	Mark	Helpful tips
10 (*a*) (i)	Hypothalamus	1	Just learn it but remember it is important in water balance as well.
10 (*a*) (ii)	Vasoconstriction, decreased.	1 (Both correct)	This is straight forward but A-type questions on this topic require you to give further explanation. In the case of vasoconstriction you would need to say that there would be reduced blood flow to the surface of the skin and so reduced heat loss by **radiation**. In the case of increased sweating you need to include increased heat loss due to **evaporation.**
10 (*b*)	Any change away from the norm / set point is detected by receptor cells which switch on a corrective mechanism. Once the condition is returned to normal / set point the corrective mechanism is switched off.	1 1	This definition of negative feedback control is usually poorly answered and is worth learning. Candidates often gain the first mark but usually miss out on the second.
10 (*c*)	Endotherms	1	A-type questions on this topic will require you to know that endotherms have homeostatic mechanism to control their body temperature and that they gain most of their body heat from their own **metabolism**. Ectotherms gain most of their body heat from their **surroundings**.
10 (*d*)	Maintains the optimum temperature for enzyme activity.	1	General statements like keeping the body warm are insufficient. When learning this topic on homeostasis you must be able to explain why the conditions need to be controlled. You should also note that any questions that involve temperature in biology usually involve an explanation in terms of enzyme action.

SECTION B (continued)

Question 11

Question	Answer(s) / separates alternative correct answers	Mark	Helpful tips
11 (*a*) (i)	As the size of the wader flock increases, the percentage of attack successes decreases. **OR** As the size of the wader flock increases, the falcons are less successful at catching pigeons. Increased vigilance. As the number of waders in the flock increases, the earlier the falcon is seen and avoided.	1 1	Note that large flocks of birds are also difficult for a falcon as their movements make it difficult for the predator to pick out an individual and attacking into a flock is dangerous. This idea could be used in the next answer but is wrong here.
11 (*a*) (ii)	The larger the flock the more difficult it is for the predator to focus on / target a single pigeon. **OR** Scattering of the large flock distracts / confuses the predator. **OR** Large flock might mob / attack the predator **OR** Greater chance of falcon being injured.	1	Many possible answers here but you need to word these carefully.
11 (*b*)	Habituation Saves energy by not responding to a harmless stimulus. **OR** Energy not wasted in an unnecessary response. **OR** Continue to feed and gain energy when stimulus is harmless	1	This is a demanding A question because of the requirement for both the term and a detailed explanation. Not enough to say 'saves energy' on its own.

SECTION B (continued)

Question 12

Question	Answer(s) / separates alternative correct answers	Mark	Helpful tips
12 (*a*)	Both axes with correct scale, labels and units. All points plotted correctly and joined with a straight line using a ruler.	1 1	This type of question often operates as an A question because of the number of tight criteria applied in the marking of it. Accurate plotting and joining is essential.
12 (*b*) (i)	D No change in mass / no net change.	1	You need to know the term isotonic to answer this one – it means an equal concentration and would therefore give no change of mass.
12 (*b*) (ii)	Turgid Flaccid / plasmolysed	1	Gaining mass means gaining water. Losing mass means losing water.
12 (*c*)	It allows a fair comparison to be made when the original / initial / starting masses are different / not the same.	1	This is a standard question. Learn up the standard answer!
12 (*d*) (i)	Volume of salt solution Same carrot / tissue / surface area. Temperature.	1 (Any two)	Always read through the description of the experiment – you won't get marks for variables already mentioned.
12 (*d*) (ii)	Repeat the experiment with more pieces of carrot.	1	Remember that repeating increases reliability. Results are then usually averaged although calculating averages itself does not increase reliability.
12 (*e*)	28% Each time the concentration decreases by 0.05 M it results in an increase in mass of 7%	1	The detail required for the justification is essential and these types of questions require you to spot the trend and use the data provided.

SECTION B (continued)

Question 13

Question	Answer(s) / separates alternative correct answers	Mark	Helpful tips
13 (*a*)	X - Pituitary gland Y - TSH / Thyroid stimulating hormone. Z – Thyroxin	2 (All correct 2, one or two correct 1)	Take care with your spelling of "pituitary". Generally credit will be given provided spelling is at least phonetically correct.
13 (*b*)	Stimulates uptake of amino acids into tissues	1	Many candidates tend to give the answer "Growth hormone causes growth", which is insufficient at this level.
13 (*c*) (i)	3.6	1	A slightly different way to ask an average – calculate the total increase in mass then divide by the number of years involved.
13 (*c*) (ii)	0-4	1	Drawing your own lines on the graph is a good idea.

SECTION B (continued)

Question 14

Question	Answer(s) / separates alternative correct answers	Mark	Helpful tips
14 (*a*)	Deaths decrease from 33 in March to 11 in June In July rises to 14 then falls to 11 in August Then increases to 35 in November	All 3 = 2 2 = 1	These questions require you to give a full description of the changes, quoting up to four or five values and units if available. Trends must be identified. Try similar questions in other papers and always quote values at appropriate points and use the exact labels and units in your answer.
14 (*b*)	They hunt on roadside verges and are usually active at night	1	The information is available in the question – reading through a couple of times is essential.
14 (*c*)	Decrease in deaths due to: Females being confined to nest / incubating eggs so less vulnerable **OR** Less deaths due to road collisions due to longer daylight hours **OR** Slight increase in July due to increase in female birds drowning when bathing,	2 (All three for 2, one or two correct for 1)	The explanations make this more demanding but again the information is there!
14 (*d*)	Drowning increases in the months just after the females emerge from the nest and are likely to be bathing.	1	Again, understanding the data is needed but quite straightforward really.
14 (*e*)	The total deaths on the graph are greater than the total on the table.	1	Looking carefully at the figures is the only way.
14 (*f*)	66.7%	1	First calculate the decrease ie. Decrease = 22 Next, express this decrease as a percentage ie $\frac{\text{Decrease}}{\text{original}} \times 100 = \frac{22}{33} \times 100 = 66.7$
14 (*g*)	6 : 2 : 1	1	Remember – no fractions.

SECTION C

Question 1A

Your answer should include / separates alternative correct answers	**Mark**	**Commentary and Helpful tips**
Fish are at a higher water concentration / hypotonic to their surroundings **OR** surroundings at lower water concentration / hypertonic to the fish.	1	**Maximum of 5 marks for this section.** This question is structured into parts to guide your response. The marks available for each part are given and these provide a clue to the length of answer required. In this question there is the same number of marks for each sub-heading and so both require equal weighting. Diagrams are clearly useful as they can save time and help to join ideas together. They must be labelled and annotated to gain any marks. For any question on water balance you should start by discussing what is meant by the term water balance and then state the particular problem faced by the organism. The arrangement document states clearly that osmoregulation in fish should include reference to the **number and size** of the glomeruli, the **filtration rate** and the **role of the chloride secretory cells** of the gills. You can see that the mark scheme allows for all of these points. Learning them as a sequence or flowchart may help you.
Water lost at gills by osmosis.	1	
Drink sea water.	1	
Few / small glomeruli **OR** aglomerular.	1	
Low filtration rate in kidney.	1	
Small volume of concentrated urine	1	
Salts secreted / removed by chloride secretory cells in the gills	1	
No sweat glands / do not sweat.	1	**Maximum of 5 marks for this section.** This section deals with water conservation in the desert rat. The arrangement document requires candidates to make the distinction between the physiological adaptations and behavioural adaptations. The behavioural adaptations include the burrowing and nocturnal habit. The other adaptations are physiological.
Low volume of urine produced.	1	
High water reabsorption in kidneys / nephrons **OR** Efficient reabsorption of water in kidneys / nephrons.	1	
Dry faeces **OR** High water reabsorption in the large intestine.	1	
Reduced water evaporation in exhaled air **OR** Water vapour in exhaled air is condensed / removed in the nasal passages.	1	
Remains cooler / avoids high water loss in daytime heat by burrowing behaviour / nocturnal behaviour	1	
Staying in burrow keeps the air moist / damp / humid.	1	

SECTION C (continued)

Question 1B

Your answer should include / separates alternative correct answers	**Mark**	**Commentary and Helpful tips**
Co-operative hunting means animals hunting in a social group / pack / team **OR** means working together in hunting / to obtain food.	1	**Maximum of 3 marks for this section.** Good definition required. The key terms to include in your answer are **'social group'**, **'working together'** and **'to obtain food'**. This is another example where candidates must use the correct terminology to gain the mark. Many candidates lose out on this mark because they do not use certain key words needed to trigger the mark. It would not be enough to say that 'larger prey can be hunted', the prey need to be **caught or killed**. Similarly, 'less energy used' on its own is not enough to get a mark. To get this mark, candidates must say that 'less energy is used **per individual'**. Have these phrases or definitions memorized.
An example of a benefit of co-operative hunting from; -larger prey can be killed / caught. -increased hunting success / more chance of catching prey. -less energy used per individual / less pursuit time per individual. -net energy gain is greater than by foraging alone. -all / weaker members / young get a share of the food.	1	
Another advantage of co-operative hunting from the above list.	1	
In dominance hierarchy there is a rank order / pecking order of individuals within a social group.	1	
A dominance hierarchy consists of dominant / alpha and subordinate individuals.	1	**Maximum of 3 marks for this section** Terms needed to describe the composition of a dominance hierarchy are dominant and subordinate. Always use the term "dominant" but an alternative is the "alpha" individual. This benefit to the dominant individual is often overlooked.
The dominant / alpha individual eats first / gets a bigger share of the food OR converse	1	
Ensures survival of the dominant individual when food is in short supply.	1	
The subordinate individuals may gain more food / energy than by foraging alone.	1	
A territory is an area which is marked / defended for feeding / hunting.	1	**Maximum of 4 marks for this section** Another good definition needed to gain the mark.
A territory ensures enough food / must contain enough food supply for mate and offspring	1	
The more food available the smaller the territory **OR** converse	1	
Holding a territory reduces competition for resources	1	
Energy is expended in marking / patrolling / defending a territory	1	Remember that any questions on foraging or obtaining food will require answers relating to energy efficiency and energy gain compared to energy loss.
Gain of energy is increased by the lack of competition **OR** foraging is made more economical	1	

SECTION C (continued)

Question 2A

Your answer should include / separates alternative correct answers	**Mark**	**Commentary and Helpful tips**
Regulator gene produces repressor molecule / protein	1	**A maximum of 5 marks for this section.** For this content the arrangement document states that the key terms which must be known include regulator gene, repressor molecule, operator, structural gene and inducer. This is another example of content which you can learn as a sequence of statements or flowchart of events. You should use the statements in points 1 to 5 to write out a flowchart showing what happens in the presence or absence of lactose.
In the absence of lactose the repressor / protein binds to the operator and the structural gene is switched-off	1	
Lactose when present binds to the repressor / protein and the operator switches-on the structural gene	1	
Lactose acts as the inducer and the enzyme is produced / synthesized.	1	
The enzyme is only produced when required **OR** Energy / cell materials are only used when necessary.	1	
Enzyme not produced/Abnormal enzyme produced due to gene mutation	1	**A maximum of 3 marks for this section.** This section of the extended response question deals with the part played by genes in controlling metabolic pathways as shown in the case of phenylketonuria. Start with the understanding that genes code for the production of the enzymes which control the chemical reactions. The order of bases in the gene determines the order of amino acids which determines the protein structure and function. A gene mutation would change the order of bases and so result in an abnormal or faulty enzyme which would then result in the events detailed in the mark scheme.
Phenylalanine cannot be metabolised as normal **OR** There is a metabolic block	1	
The compounds / phenylpyruvic acid / high concentration of phenylalanine produced affect the developing brain / causes mental retardation.	1	
Babies blood tested **OR** Affected individuals follow a low phenylalanine diet.	1	

SECTION C (continued)

Question 2A (continued)

Your answer should include / separates alternative correct answers	**Mark**	**Commentary and Helpful tips**
Coherence The writing **should** be under **sub headings** or divided into **paragraphs**. A sub heading / paragraph for Jacob-Monod and a sub heading/paragraph for PKU. The account **must** be presented in a **logical** and **progressive** way. Related information must be grouped together. Information on Jacob-Monod (3 points) and information on PKU together (2 points). Five points needed.	1	These criteria are usually well understood by candidates and are fairly easy to follow. The fact that you need five points to trigger this mark usually means that it is the better candidates who get it.
Relevance Must not give details not directly related to Jacob-Monod and PKU. Must give at least 3 relevant points on Jacob-Monod and 2 points on PKU. Five points needed. **Both must apply correctly to gain the Relevance Mark.**	1	

SECTION C (continued)

Question 2B

Your answer should include / separates alternative correct answers	**Mark**	**Commentary and Helpful tips**
IAA increases plasticity / elasticity of cell wall or ability of cell wall to stretch	1	**A maximum of 5 marks for this section.** Very few candidates are able to give this answer which explains how the cells can undergo elongation when the vacuole forms and stretches them.
IAA leads to elongation of cells at shoot / root tip and gives growth in length.	1	
IAA inhibits development / growth of the lateral buds and promotes growth of the apical / terminal bud.	1	Be able to give the term apical dominance but also to describe it as detailed.
As IAA concentration decreases in a leaf an abscission layer forms and the leaf falls off.	1	This functions as an A- type question because although candidates make the link between the IAA and the term leaf abscission, they do not show that they understand that it is the **low concentration** which causes leaf abscission. Try to remember it as a fall in IAA causes leaf fall.
IAA causes ovary (wall) to develop into (wall of) fruit.	1	Fruit formation in more detail.
In directional light IAA accumulates on the dark side / side away from the light and this leads to growth towards the light / phototropic response.	1	The term phototropism as a **growth** movement towards light should be known. DO NOT say that they 'bend' towards the light. You must also include detail of the uneven distribution and concentration of IAA causing greater elongation and growth on the dark side.
The embryo produces gibberellic acid (GA)	1	**A maximum of 3 marks for this section.**
Gibberellic acid (GA) diffuses out to the aleurone layer.	1	
GA stimulates cells of the aleurone layer to synthesise the amylase.	1	

SECTION C (continued)

Question 2B (continued)

Your answer should include / separates alternative correct answers	**Mark**	**Commentary and Helpful tips**
Coherence The writing must be under sub headings or divided into paragraphs. A sub-heading / paragraph for IAA and a sub heading / paragraph for GA. Related information must be grouped together. The account must be presented in a logical and progressive way. Information on IAA together (3 points) and information on GA together (2 points).	1	These criteria are usually well understood by candidates and are fairly easy to follow. The fact that you need five points to trigger this mark usually means that it is the better candidates who get it.
Five points needed		
Relevance Must not give detail not directly related to IAA and GA. Must give at least 3 relevant points on IAA and 2 points on GA. Five points needed. **Both must apply correctly to gain the Relevance Mark.**	1	

EXAM 3 **WORKED ANSWERS**

SECTION A

Question	Mark	Helpful tips
1	C	As soon as you see the term 'active', make the link to the need for energy in the form of ATP and so respiration. So in this question the connection is mitochondria.
2	D	Plant cell walls have cellulose **fibres** and are **fully** permeable.
3	D	Learn the term hypotonic in terms of relative water concentration. Hypotonic really means "higher water concentration than". Pure water is the most hypotonic of all!
4	A	Light is needed to convert RuBP to GP in the Calvin cycle. No light – no conversion so more RuBP and less GP
5	C	Membrane flexibility is needed to enable phagocytes to surround and engulf particles to enclose them in a food vacuole.
6	A	Foreign material like a transplant acts as an antigen which causes the lymphocytes to **gen**erate **anti**bodies.
7	A	ATP transfers energy and NADP carries the hydrogen needed to reduce the CO_2 in carbon fixation.
8	D	Once the phagocytes have surrounded and engulfed the bacteria inside a food vacuole, the lysosomes fuse with the membrane and release their enzymes inside to digest them.
9	B	Look for the lowest number that will divide into both values given and go from there.
10	B	Since contraction still occurred after the ATP had been boiled, this proves that the ATP is not made of protein and is not denatured. The others are possible!
11	B	Since it is caused by a dominant allele and the father is heterozygous he has a 50% or 1 **in** 2 chance. The normal allele comes from the mother
12	D	A father never passes his X chromosome to a son, only his Y chromosome. This means that answers A or B are not correct. His X chromosome will, however, be passed on to all his daughters and so 100% will inherit this recessive allele.
13	D	In polyploidy an individual possesses one or more sets of chromosomes in excess of the normal diploid number. This results in improved vigour eg. increased growth, crop yield, fruit size or disease resistance.
14	B	Non-disjunction involves the failure of the spindle to separate pairs of chromosomes during cell division, resulting in a change in the number of chromosomes in the daughter cells.
15	C	In genetic engineering the endonuclease 'cuts' and ligase 'seals'. In somatic fusion the enzyme cellulase removes the cell walls.

SECTION A (continued)

Question	Mark	Helpful tips
16	C	Somatic fusion overcomes sexual incompatability between different plant species. This is a good definition to learn.
17	A	Good sequence to learn.
18	D	A short term learned response in which an organism learns not to respond to a harmless stimulus, thereby saving energy.
19	C	First step is to calculate the average. (120 divided by 5 = 24). Then calculate 75% more. ($24 \times 0.75=18$) then add it. This comes to 42. Tricky!
20	C	First step is to calculate the percentage inhibition for the grasses from both areas. (1) 0.3 divided by $3.0 \times 100 = 10\%$ and (2) 2.1 divided by $2.8 \times 100 = 75\%$. The difference is therefore 65%.
21	B	1. Cambium in secondary thickening 2. Apical meristems produce IAA 3. Animals do not have meristems and the cell division which occurs in meristems is mitosis.
22	C	Remember **ROSE**: **R**egulator, **O**perator, **S**tructural, **E**nzyme. As a result of the mutation it means that the structural gene is never switched off and the enzyme is made continuously.
23	D	A gene mutation results in the enzyme that converts phenylalanine into tyrosine not being produced.
24	A	Endotherms have a physiological internal temperature control mechanism and derive most of their heat from this.
25	A	IAA can be used as a selective weedkiller, rooting powder for cuttings, to prevent fruit fall or to produce seedless fruits.
26	D	Good sequence to learn
27	A	The hairs flatten and reduce the insulating effect of the air they can trap and vasodilation increases blood flow to the surface of the skin causing increased heat loss by radiation.
28	B	The decrease is $80 - 60 = 20$. Expressed as a percentage this is 20 divided by $80 \times 100 = 25\%$
29	B	Since compound R cannot be converted to compound S this shows that the mutation has occurred in the gene coding for enzyme 2
30	C	At week 6 the reading is 500 to 200 which gives the ratio of 2.5:1

SECTION B

Question 1

Question	Answer(s) / separates alternative correct answers	Mark	Helpful tips
1 (*a*) (i)	400 – 420 nm	1	Selecting information questions require care. In this question, first draw a line across the graph at 40% on the Y-axis. Then draw a line down onto the X-axis at each point it crossed the graph. Use a ruler to do this.
1 (*a*) (ii)	Even at wavelengths when the absorption by chlorophyll a is low the rate of photosynthesis remains high **OR** At wavelengths of between 450 and 500 nm when the absorption by chlorophyll a is low the rate of photosynthesis remains high	1	Seems fairly straight forward but is often poorly answered. Whenever possible try to use evidence and data/values to support your answer. Always use the phrase or statement given in the stem of the question or on the axes labels or table headings in your answer.
1 (*a*) (iii)	They broaden the absorption spectrum so that more energy is available for photosynthesis.	1	Good definition to learn for the role of the accessory pigments. Remember that they then pass this energy onto the chlorophyll a.
1 (*b*) (i)	Reflected or transmitted.	1	Remember ART: Absorbed, Reflected and Transmitted.
1 (*b*) (ii)	Granum	1	Think of these as stacks of membrane trays on which the pigments are spread out to provide a large surface area to absorb light energy.

SECTION B (continued)

Question 2

Question	Answer(s) / separates alternative correct answers	Mark	Helpful tips
2 (*a*)	ATP	1	ATP is the link between the energy producing reactions (respiration) and the energy requiring processes (eg. mitosis, DNA replication. protein synthesis etc)
2 (*b*)	NAD	1	When asked for the name of the hydrogen carrier, just use NAD and do not give it in its reduced form NADH. To avoid confusion with the photosynthesis hydrogen carrier remember NAD for respiration and NAD**P for Photosynthesis**.
2 (*c*)	Acts as the final hydrogen acceptor	1	Oxygen joins with the hydrogen and forms water
2 (*d*)	Aerobic respiration produces 38ATP whereas anaerobic respiration only produces 2 ATP	1	The 'less efficient' refers to the anaerobic process not releasing as much energy from glucose as the aerobic process.
2 (*e*)	1 Lactic acid 2 Ethanol **AND** carbon dioxide	1 1	Lactic acid on its own in muscles but **both** ethanol and carbon dioxide for plants
2 (*f*)	Cristae of the mitochondria	1	These are the folds of the mitochondria providing an increased surface area for the cytochrome system.
2 (*g*)	Glycolysis	1	The splitting of glucose, which although needing some ATP to start the reaction produces an overall or net gain of 2 ATP.

SECTION B (continued)

Question 3

Question	Answer(s) / separates alternative correct answers	Mark	Helpful tips
3 (*a*) (i)	1 Deoxyribose 2 Phosphate	1	Straight forward labeling of these two parts of a DNA nucleotide. Made a little harder by needing two answers for one mark
3 (*a*) (ii)	3 Guanine 4 Cytosine 5 Adenine	2	Testing your knowledge of DNA base pairs. **A**nthea **T**urner and **G**eorge **C**loony. Take your time and match up the bases correctly.
3 (*a*) (iii)	Hydrogen	1	The weak hydrogen bond between the base pairs which makes the separation of the two DNA strands (unzipping) much easier.
3 (*a*) (iv)	ATP **OR** enzymes	1	The other two are already shown in the diagram. (The DNA which acts as a template and the supply of free DNA nucleotides)
3 (*b*)	So that each daughter cell has all the (genetic) information it needs to carry out all of its chemical reactions/activities/functions	1	Useful phrase to learn for this question. Remember that each gene codes for a specific protein eg. enzyme that controls a chemical reaction.
3 (*c*)	Gene	1	Section of DNA with a specific sequence of bases coding for a specific protein.
3 (*d*)	Type of protein - Globular Function of endonuclease - used in genetic engineering to cut (useful or required) gene from chromosome **OR** cuts DNA into fragments **OR** cuts plasmid open Example of fibrous protein - collagen	2 (all three boxes = 2, 2 or 1 box = 1 mark)	Fibrous and globular are the only types of protein required at Higher. Collagen is the only named fibrous protein that can be asked about.

SECTION B (continued)

Question 4

Question	Answer(s) / separates alternative correct answers	Mark	Helpful tips
4 (*a*) (i)	X = Protein coat Y = Nucleic acid/DNA/RNA	1	Straightforward but remember viruses can have either DNA or RNA
4 (*a*) (ii)	ATP/amino acids/nucleotides	1 (Any 2)	ATP for energy. Amino acids for the synthesis of the viral protein coat. Nucleotides for the replication of the viral nucleic acid.
4 (*b*) (i)	Acts as a barrier to prevent the spread of infection/bacteria/fungi/viruses to other parts of the plant	1	Most candidates stop at 'barrier". You need to go on to include the second part of the answer. Remember! At Higher, be safe – Explain.
4 (*b*) (ii)	Tannin **OR** nicotine **OR** cyanide (any 2)	1	**T**hree **N**asty **C**hemicals = Tannin, Nicotine and Cyanide

Question 5

Question	Answer(s) / separates alternative correct answers	Mark	Helpful tips
5 (*a*)	1 DNA/chromosomes replicate	3 (One mark for each)	Good understanding of meiosis needed here! DNA replication takes place followed by the first then second meiotic divisions.
5 (*b*)	2 Homologous pairs of chromosomes are separated/pulled apart		
5 (*c*)	3 Chromatids are separated/pulled apart		
5 (*b*) (i)	Chiasma/chiasmata	1	A touching moment between the homologous pairs of chromosomes before they line up along the equator to be pulled apart at the first meiotic division.
5 (*b*) (ii)	Term: Crossing over Explanation: Linked genes are separated producing recombinant gametes/new allele combinations	1 1	Many students learn crossing over and independent assortment of chromosomes as the two sources of variation that take place during meiosis However, they are not so good at learning the language to explain how they increase variation.

SECTION B (continued)

Question 6

Question	Answer(s) / separates alternative correct answers	Mark	Helpful tips
6 (*a*)	As the length of time exposed to radiation increases the estimated number of mutations per cell increases (Exposure to) Source 2 radiation causes more mutations per cell (than Source 1)	1 1	Examine the results and trends. Remember to use the language and terms given to you in the question.
6 (*b*)	100%	1	With percentage increase questions always work out the increase first and then express it as a percentage. Increase = 20 % = increase/original value × 100 20/20 × 100 = 100
6 (*c*)	Prediction: 67 Justification: The increases are doubling every hour	1	When predicting from a table of results, look for a pattern in the values either increasing or decreasing by a certain factor and then apply this observation to obtain your prediction.

SECTION B (continued)

Question 7

Question	Answer(s) / separates alternative correct answers	Mark	Helpful tips
7 (*a*) (i)	76	1	A more difficult question which involves a few steps to get the answer. First step is to identify the volume of water being drunk at the start = 19 cm^3. We now multiply the volume (19) by 2 for a 2 kg salmon and then another 2 for the 2 hour period. $19 \times 2 \times 2 = 76$
7 (*a*) (ii)	Increases Increases	1 (Both correct for one mark)	These will produce large volumes of dilute urine needed to restore the water balance.
7 (*a*) (iii)	Actively pump out/excrete salt	1	In the sea, the salmon will need to drink saltwater to replace the water it is losing. The chloride secretory cells need to use energy to pump the excess salts out.
7 (*b*)	Behavioural: Nocturnal or hunts or is active at night **OR** Burrowing habit Physiological:Absence of sweat glands/does not sweat **OR** Dry mouth/nasal passages **OR** High/efficient reabsorption of water in large intestine **OR** High/efficient reabsorption of water by kidneys **OR** Produces dry faeces	1 1	Need to learn the terms behavioural and physiological adaptations for more difficult questions or extended response questions on this topic. More difficult questions would also need you to explain how the adaptations have their effect.

SECTION B (continued)

Question 8

Question	Answer(s) / separates alternative correct answers	Mark	Helpful tips
8 (*a*) (i)	Ensure the same number of leaves on each shoot **OR** Ensure the surface area of each shoot is about the same	1	Water loss occurs through the leaves so they must be similar in number and size to allow valid comparison.
8 (*a*) (ii)	To prevent evaporation of water from the flask to ensure that any change in mass was due to transpiration	1	The added explanation is needed for this A mark.
8 (*a*) (iii)	Repeat the investigation with more shoots from more plants of both species	1	It is no good simply saying repeat the experiment - at Higher more detail is required.
8 (*a*) (iv)	Light intensity/humidity/temperature/air movements	1 (Any two)	Note that an increase in all of these will increase transpiration except an increase in humidity which decreases it.
8 (*b*)	Scales enclosed including zeros and labels directly from data table **AND** Points accurately plotted and joined with a straight line	1 1	Take care – many candidates lose needless marks here. Zeros are needed on both scales, there must be labels and units on the axes. Plots must be accurate and joined using a ruler.
8 (*c*)	3 g per hour	1	A different way of asking an average. Total loss is 12 g divided by 4 gives 3 g per hour.
8 (*d*)	Cooling effect **OR** water for turgidity **OR** water for photosynthesis/photolysis **OR** supplies plant cells with minerals/nutrient ions	1	Straight forward but often not remembered well.

SECTION B (continued)

Question 9

Question	Answer(s) / separates alternative correct answers	Mark	Helpful tips
9 (*a*) (i)	Food sources close to the hive result in a round dance while sources at a greater distance result in a waggle dance **OR** A round dance is performed when the food source is within 100 m of the hive and a waggle dance is performed at distances greater than 100 m	1	Information from the diagram and table.
9 (*a*) (ii)	X = waggle dance Y = round dance Justification: Y lies as close as sources 3 and 6 which both result in a round dance. X lies about as far away as 8 which results in a waggle dance.	1 1	Justification makes this more difficult.
9 (*a*) (iii)	1 As the distance to the food sources increase the average wing vibration period increases 2 As the distance to the food sources increase the number of dance runs decreases rapidly then slowly	1 1	1 Easy enough 2 Watch for the changing shape of the graph line here.
9 (*a*) (iv)	1.8 8	1	First part is straight forward selecting information from the graph.
9 (*a*) (v)	6.4	1	Dance numbers seem to have levelled out – no reason to believe that they would now change. However, you need to have noticed that the measurements were for every 15 seconds and the question asks for the number of dances per minute. You now need to multiply by 4, ie. $1{\cdot}6 \times 4 = 6{\cdot}4$
9 (*b*)	Behaviour gives /ensures optimal foraging **OR** Maximizes energy gain **OR** Prevents other bees wasting energy searching/foraging/locating food	1	Remember that in economic foraging the energy used in foraging must be less than the energy gained in food.

SECTION B (continued)

Question 10

Question	Answer(s) / separates alternative correct answers	Mark	Helpful tips
10 (*a*) (i)	As the flower density increases the territory size decreases but flower density greater than 7 makes no further difference to territory size **OR** The territory size decreases as the flower density increases	1	Be careful that you express these relationships in the correct order. You would not be awarded the mark if you had said that '**As** the territory size increases the flower density decreases' because this makes it look as if it is the large territory that has caused the low flower density.
10 (*a*) (ii)	Intraspecific competition in this species will be very high where the flowers are densest Where flowers are least dense the size of territory has to be very large and so difficult to defend	1 1	Note that there are problems on both sides of the preferred flower density! Very attractive territories with high density of flowers may be difficult to compete for.
10 (*a*) (iii)	The energy lost/used/expended defending the large territory would be greater than the energy gained from the low flower density	1	Most questions on this area require answers in terms of energy!
10 (*b*)	Interspecific	1	Inter = between **different** species (Remember international games are between different countries). Intra = between the same members of a species.
10 (*c*)	Conserving energy by only using the energy required to cope with the threat and maintain the territory. **OR** Only enough energy needs to be used to achieve the defence of the territory/cope with the level of threat	1	Again making the link between the behaviour and the energy expenditure.

SECTION B (continued)

Question 11

Question	Answer(s) / separates alternative correct answers	Mark	Helpful tips
11 (*a*) (i)	10^{-3} to 10^{-1}	1	Just take care to select the information carefully
11 (*a*) (ii)	10^{-2}	1	Note where both of the graphs cross or intersect.
11 (*a*) (iii)	As concentration of IAA increases from 0 to 10^{-4} the percentage stimulation increases from 0 to 50%. As the concentration increases to 10^{-1} the percentage stimulation drops to zero. Growth is then inhibited and reaches 100% inhibition at 10^{1}	2 (All three points for 2 marks, 2 or 1 point for 1 mark)	In these questions which ask you to describe using values, you must be ready to give a really full and detailed response. Use at least 3–5 values using the terms and units from both axes.
11 (*b*)	IAA alone: Prevention of leaf abscission, fruit formation, phototropism GA alone: amylase production in barley grains, breaking of bud dormancy IAA and GA: Elongation of internode cells	3 (All 6 rows correct for 3 marks; 5 or 4 rows correct for 2 marks; 3 or 2 rows for 1 mark	Good knowledge of the role of these plant growth substances needed here.

Question 12

Question	Answer(s) / separates alternative correct answers	Mark	Helpful tips
12 (*a*)	Goldenrod Flowering occurs when the number of hours of light is below a certain critical value **OR** Flowering occurs when the number of hours of darkness is above a certain critical value	1	Make sure that you have learned the definitions for long and short-day plants in the form given in this answer.
12 (*b*)	Onset of breeding/courtship/mating/nest building **OR** migration	1	Photoperiod is the number of hours of light in a day. Photoperiodism is the term used to describe the response of an organism to changes in the photoperiod.

SECTION B (continued)

Question 13

Question	Answer(s) / separates alternative correct answers	Mark	Helpful tips
13 (*a*) (i)	Pancreas	1	**PIG** = **P**ancreas, **I**nsulin and **G**lucagon
13 (*a*) (ii)	Insulin concentration increases Glucagon concentration decreases	1	More difficult questions and extended response questions will require you to know that insulin increases the permeability of the liver and muscle cells to allow greater uptake of glucose into these cells.
13 (*a*) (iii)	Glycogen Liver both	1	Note that in these questions because of the similarity of the word glucagon and glycogen that the spelling needs to be correct.
13 (*b*)	Any change away from the norm is detected by receptor cells which switch on a corrective mechanism. Once the conditions are returned to normal the corrective mechanism is switched off.	1 1	A really good definition to learn for negative feedback control. Most candidates stop after the first sentence and lose out on the second mark.

Nlg feedback

Question 14

Question	Answer(s) / separates alternative correct answers	Mark	Helpful tips
14 (*a*) (i)	210	1	250 – 40 = 210 Note the factor of ×100 on the Y axis that has to be applied.
14 (*a*) (ii)	Captive breeding/reserves **OR** Culling grey squirrels	1	Suitable suggestion needed here. Note that in these questions on populations that the monitoring of populations and obtaining data on its own is not enough. You need to be able to say what the next steps or action to be taken would be.
14 (*b*)	Indicators of pollution **OR** management of raw materials **OR** food supplies **OR** control of pest species	1 (Any two)	Remember **POL RAWCONPEST** **POL** = Pollution **RAW** = Raw materials **CON** = Conservation of endangered species **PEST** = Pest species numbers

SECTION C

Question 1A

Your answer should include / separates alternative correct answers	**Mark**	**Commentary and Helpful tips**
Glycolysis occurs in the cytoplasm.	1	**A maximum of 5 marks for this section.** The first stage of respiration.
Glucose is broken down/converted/oxidised to (two molecules of) pyruvic acid/pyruvate.	1	Glyco = glucose. Lysis =. to split Remember that the liver stores extra glucose as a substance called glycogen.
C6 compound broken down to 2 × C3.	1	You need to know the number of carbon atoms for each compound.
Step by step breakdown by enzymes **OR** series of enzyme controlled reactions.	1	If this mark is awarded here, then it cannot be awarded again at point 16.
Net gain/net production of 2 ATP or explanation of net gain of ATP.	1	Net = overall. You must be able to explain this in terms of needing 2 ATP to start glycolysis but gaining 4 ATP in return and therefore making an overall gain of 2 ATP.
NAD accepts hydrogen/NADH produced **and** is transferred to the cytochrome system or cristae/electron transfer system/hydrogen transfer system.	1	NAD is the hydrogen carrier in respiration. This is an A type question because you need to state where the hydrogen is then transported. If this mark is awarded here, then it cannot be awarded again at point 15.
Oxygen is not required or anaerobic or occurs in both aerobic or anaerobic conditions	1	If this mark is awarded here, then it cannot be awarded again at point 9.
Krebs cycle occurs in the matrix of the mitochondrion	1	**A maximum of 5 marks for this section.** The second stage of respiration. Krebs cycle starts the aerobic phase of respiration. The matrix is the fluid filled centre of the mitochondrion which contains the enzymes that control this cycle of chemical reactions.

SECTION C (continued)

Question 1A (continued)

Your answer should include / separates alternative correct answers	**Mark**	**Commentary and Helpful tips**
It requires oxygen or this is the aerobic phase	1	This mark will only be awarded if point 7 has not been given already.
A C2 acetyl group is produced from the pyruvic acid	1	Many candidates miss this step out and therefore lose this mark.
The acetyl group then joins with coenzyme A (CoA) to form acetyl-CoA.	1	Many candidates miss this step out and therefore lose this mark.
The acetyl-CoA reacts/joins/combines with a C4 compound to form a C6 compound/citric acid.	1	The 4 compound names you are required to know for respiration are GPAC = Glucose, Pyruvic, Acetyl CoA and Citric acid.
A (cyclical) series of reactions takes place back to the C4 compound or/shown by a diagram or drawing	1	Citric acid is gradually converted back to the 4-carbon compound. A good well labelled and arrowed diagram including the names of the stages, intermediate compounds and carbon numbers will gain many of these marks.
Carbon dioxide is produced/released	1	The carbons are lost in the form of carbon dioxide.
NAD accepts hydrogen or NADH/$NADH_2$ is produced **AND** is transferred to the cytochrome system/cristae/ electron transfer system.	1	NAD is the hydrogen carrier in respiration. This is an A type question because you need to state where the hydrogen is then transported. This mark is only awarded here if it has not already been awarded at point 6.
(Krebs cycle) is controlled by enzymes or Krebs cycle needs enzymes or processes are controlled by enzymes. (Only awarded if point 4 has not been awarded in the glycolysis section.	1	This mark is only awarded here if it has not already been awarded at point 4. Although really a C type mark candidates often forget to include the role of enzymes in these extended response questions.

SECTION C (continued)

Question 1B

Your answer should include / separates alternative correct answers	**Mark**	**Commentary and Helpful tips**
mRNA is single stranded.	1	**Maximum of 3 marks from points 1 to 5.** You must write out the full names of the bases. Marks will not be awarded for the letters on their own. Remember that the base uracil replaces thymine which is present in DNA.
It is made of (RNA) nucleotides	1	
Each nucleotide has a base, ribose and a phosphate	1	
The bases are guanine, cytosine, adenine and uracil. (not letters A,U,G,C).	1	
The nucleotides are connected by the sugar-phosphate backbone/ bonding.	1	
mRNA carries information or code of bases from the nucleus	1	**Maximum of 7 marks from points 6 to 14.** mRNA carries a copy of the order of the bases from the gene out of the nucleus into the cytoplasm and onto the ribosomes. It is quite useful to learn the facts in this section as a sequence or flowchart. Once again a good well labelled diagram may focus your thoughts and secure many of the marks.
The mRNA attaches to the ribosome **OR** the mRNA goes to/moves along the ribosome	1	
Each group of three bases on the mRNA strand is called a codon	1	
tRNA molecules transport amino acids to the ribosomes.	1	
tRNA transports specific amino acids	1	
The three exposed bases on the tRNA is called an anticodon.	1	
The anticodons of the tRNA match/pair (with complementary) codons of the mRNA	1	
The correct amino acid is added/joined onto the growing protein/polypeptide.	1	
The sequence of bases/codons on the mRNA determines the sequence of amino acids	1	

SECTION C (continued)

Question 2A

Your answer should include / separates alternative correct answers	**Mark**	**Commentary and Helpful tips**
Nitrogen is required for protein/amino acids	1	**A maximum of 5 marks for this section.** Remember **N**itrogen for ami**N**o acid synthesis, for protei**N**s and **N**ucleic acids. Having gained these marks you must now go on to explain the role of the proteins and the DNA or RNA to get these A type marks detailed in points 2 and 4. In the same way, candidates may remember that Nitrogen is needed to make chlorophyll but do not go on to state its role in photosynthesis.
Protein required for enzymes/membranes	1	
Nitrogen required for DNA/RNA/nucleotides/nucleic acids	1	
DNA/RNA essential for mitosis and cell division/for protein synthesis	1	
Nitrogen required for chlorophyll	1	
Chlorophyll for photosynthesis	1	
Lacking nitrogen – reduced shoot growth **OR** chlorosis **OR** pale green or yellow leaves **OR** red leaf bases **OR** longer root system	1	
Vitamin D required for uptake of calcium from intestines. **OR** Vitamin D required for calcium into bone.	1	**A maximum of 3 marks for this section.** Not difficult but often forgotten.
Deficiency leads to rickets/soft bones	1	
Iron needed for haemoglobin/many enzymes/hydrogen carrying systems	1	Candidates usually remember the role of iron in haemoglobin and anaemia but are less likely to remember its other roles in some enzymes and the cytochrome system.
Deficiency leads to anaemia	1	
The writing must be under sub headings or divided into paragraphs. A sub-heading / paragraph for the importance of nitrogen and a sub heading / paragraph for Vitamin D and iron in animals. Related information must be grouped together. The account must be presented in a logical and progressive way. Information on the importance of nitrogen together (3 points) and information on Vitamin D and iron together (2 points). Five points needed	1	These criteria are usually well understood by candidates and are fairly easy to follow. The fact that you need five points to trigger this mark usually means that it is the better candidates who get it.
Must not give details of any other plant minerals or other environmental influences such as lead or drugs. Five points needed. Must give at least 3 relevant points on the importance of nitrogen and 2 points on Vitamin D and iron. **Both must apply correctly to gain the Relevance Mark.**	1	

SECTION C (continued)

Question 2B

Your answer should include / separates alternative correct answers	**Mark**	**Commentary and Helpful tips**
As the population density increases, density dependent factors have a greater effect or converse	1	**A maximum of 4 marks for this section.** A good definition to learn. Many candidates do not fully understand this effect and do not express themselves properly and so lose the mark.
Density dependent factors include: Disease/parasites, food supply, predation and competition for space, food or habitat.	1 (Any two for one mark)	Straight from the arrangement document. Be careful not to give the answer 'competition' on its own, it has to be competition for food or space or habitat.
When population density increases then disease/parasites spread more easily **OR** When population density increases the food supply decreases **OR** When population density increases then predation increases **OR** When population density increases then competition for space, food or habitat increase.	2 (Any 2 for one mark each)	This sometimes functions as a more difficult mark because candidates do not describe these relationships as detailed in the mark scheme. It is essential to use the words increase and decrease.
The effect of the increase in the effect of the factor is to reduce the population	1	
Effect tends to return the population to a stable size/carrying capacity/ size that the environment can support	1	This mark is often overlooked or forgotten or perhaps has not been emphasised.
Succession is the sequence of plant communities inhabiting an area. **OR** Successsion is the gradual change in the species of plants present in a particular habitat.	1	**A maximum of 4 marks for this section.** Good definition to learn.

SECTION C (continued)

Question 2B (continued)

Your answer should include / separates alternative correct answers	**Mark**	**Commentary and Helpful tips**
Succession is unidirectional	1	The changes go in one direction from the pioneer community then the intermediate community through to the final climax community.
The communities/populations/plants modify/change the habitat/ increase soil fertility, making it more suitable for other new communities/populations/plants	1	This requires you to describe the changes which take place and to understand that the new conditions result in new communities.
Characteristics of the climax community are that it has; a greater species diversity **OR** a greater biomass, **OR** more complex food webs.	1 (Any two for one mark)	Learn these 3 characteristics of a climax community.
The final community is the climax community.	1	This section on plant succession can be shown in the form of a well labelled and annotated diagram detailing the information in points 7 to 11.
The writing must be under sub headings or divided into paragraphs. A sub-heading / paragraph for density dependent factors and a sub heading / paragraph for plant succession. Related information must be grouped together. The account must be presented in a logical and progressive way. Information on density dependent factors together and information on plant succession together. Five points needed. At least 3 from from either section and 2 from the other.	1	These criteria are usually well understood by candidates and are fairly easy to follow. The fact that you need five points to trigger this mark usually means that it is the better candidates who get it.
Must not give detail on density independent factors. Five points needed. At least 3 from from either section and 2 from the other. **Both must apply correctly to gain the Relevance Mark.**	1